Applied Nonparametric
Statistical Methods

OTHER STATISTICS TEXTS FROM
CHAPMAN AND HALL

Applied Nonparametric Statistical Methods

Peter Sprent

Emeritus Professor of Statistics
University of Dundee, Scotland, UK

London   New York
CHAPMAN AND HALL

First published in 1989 by Chapman and Hall Ltd
11 New Fetter Lane, London EC4P 4EE
Published in the USA by Chapman and Hall
29 West 35th Street, New York NY 10001

© 1989 P. Sprent

Typeset in 10/12 Times by
Thomson Press (India) Ltd, New Delhi

Printed in Great Britain by
J. W. Arrowsmith Ltd, Bristol

ISBN 0 412 30600 X (hardback)
      0 412 30610 7 (paperback)

British Library Cataloguing in Publication Data

Sprent, Peter
    Applied nonparametric statistical
    methods.
    1. Nonparametric statistical mathematics.
    Applications
    I. Title
    519.5

ISBN 0-412-71600-X

Library of Congress Cataloging in Publication Data

Sprent, Peter
    Applied nonparametric statistical methods/Peter Sprent.
        p. cm.
    Bibliography: p.
    Includes index.
    ISBN 0-412-30600-X. ISBN 0-412-30610-7 (pbk.)
    1. Nonparametric statistics. I. Title.
QA278.8.S74 1989
519.5--dc19

Contents

Preface

This book is a practical introduction to statistical techniques called **nonparametric methods**. Using examples, we explain assumptions and demonstrate procedures; theory is kept to a minimum. We show how basic problems are tackled and try to clear up common misapprehensions so as to help both students of statistics meeting the methods for the first time and workers in other fields faced with data needing simple but informative analysis.

An analogy between experimenters and car drivers describes our aim. Statistical analyses may be done by following a set of rules without understanding their logical basis, but this has dangers. It is like driving a car with no inkling of how the internal combustion engine, the gears, the ignition system, the brakes actually work. Understanding the rudiments helps one get better performance and makes driving safer; appropriate gear changes become a way to reduce engine stress, prolong engine life, improve fuel economy, minimize wear on brake linings. Knowing how to change the engine oil or replace worn sparking plugs is not essential for a driver, but it will reduce costs. Learning such basics will not make one a fully fledged mechanic, even less an automotive engineer; but it all contributes to more economical and safer driving, alerting one to the dangers of bald tyres, a leaking exhaust, worn brake linings.

Many research workers, industrialists and businessmen carry out their own basic statistical analyses. Professional statisticians may deplore this (as skilled mechanics grumble about do-it-yourself car servicing). These professional attitudes are a mixture of self-interest and genuine concern that serious mistakes may be made by the amateur.

This book is not meant to turn those meeting data in their daily work into professional statisticians, any more than a guide to do-it-yourself car servicing will turn one into a trained mechanic.

Relatively straightforward nonparametric counterparts of old established statistical tools are dealt with in Chapters 1 to 8 with only occasional references to sophisticated material (e.g. log-linear models in Section 8.3).

In Chapter 9 we look at some recent developments, while Chapter 10 outlines more advanced techniques; use of these will generally require guidance from a professional statistician, but it is handy for data-producers to know what is on offer. Using our motoring analogy, few do-it-yourself car servicers have the skill or tools to replace the gearbox or install a new engine; but it helps to know a little about available alternatives which may include removal and repair, replacement by a new unit or by a reconditioned one; all have pros and cons; so too with advanced statistical analyses.

We use real (or at least realistic) data in examples and exercises; some specially obtained for this book, some extracted from larger published sets with sources indicated. Reference to the source will often show that the complete data sets have been analysed with different objectives using more advanced techniques (parametric or nonparametric). These advanced analyses are akin to specialist mechanical maintenance of a car.

The book is a detailed and modernized development from an earlier work of mine, *Quick Statistics* (Penguin Books, 1981). Emphasis there was on the 'quick'; here it is on the 'statistics'. Some common ground is covered but the change in emphasis is a logical development in the light of new attitudes to statistical methods stimulated by availability of ever-increasing computer power.

To keep the book at a reasonable length for an introductory text without making discussion of each topic too terse, I have given references to accounts of some topics that were strong candidates for inclusion where these are well covered at this level by other writers.

P. Sprent
December 1987

1

Introducing nonparametric methods

1.1 BASIC STATISTICS

In most of this book we assume only a rudimentary knowledge of statistics like that provided by a service or introductory course of 10 to 20 lectures, or by reading a simple text like *Statistics Without Tears* (Rowntree, 1981).

Readers knowing no statistics may still follow this book by reading a text like Rowntree's in parallel; those happy with elementary mathematical notations may prefer the more sophisticated but basic approach in Chapters 1 to 8 of *Statistics for Technology* (Chatfield, 1983), or one of the many other introductory texts by both British and American authors.

In the Appendix we summarize some general statistical concepts that are especially relevant to nonparametric methods. An 'A' before a section number implies it is in the Appendix and references to the Appendix are given in the form 'see Section A6', etc. Section A8 gives tables for nonparametric procedures. The headings of these tables have an 'A' before their number, e.g. Table A6 is the sixth table in Section A8.

Basic statistics courses do not always include practical applications of nonparametric methods. In this chapter we survey some fundamentals and give one or two illustrations. Specific techniques are discussed in later chapters.

1.1.1 Parametric methods

Before explaining the nonparametric kind, a word about parametric inference.

Early in statistics courses one meets random variables belonging to the family of **normal distributions**. Members of that family are distinguished by different means and/or variances, often denoted by the Greek letters μ, σ^2 respectively, and called **parameters.**

Another well known family is that of **binomial distributions**, characterized by two parameters, n, the number of observations and p, the probability of one of two possible outcomes at each observation (often called success and failure). The number of successes in a sequence of n independent observations (trials) when there is probability p of success at each has a binomial distribution.

Given one or more sets of independent observations (called random samples) assumed to come from some distribution belonging to a named family, we often want to infer something about the unknown parameters. The sample mean provides an estimate of the distribution (or population) mean. With a sample from a normal distribution the t-test (Section A4) may be used to decide if the sample is consistent with an *a priori* hypothesized population mean μ_0. The related t-distribution lets us establish a **confidence interval**: an interval (see Section 1.3.1) in which we are reasonably confident the true but unknown mean μ lies.

For a binomial distribution, if there are r successes in n independent observations we call r/n a **point estimate** of p and we may test whether that estimate supports an *a priori* hypothesized value p_0, say, or obtain a confidence interval for the true value of p, the probability of success at each independent observation.

In practice an assumption that observations come from a particular family of distributions such as the normal or the binomial may be quite reasonable. Experience, backed to some extent by theory, suggests that, for many measurements, inferences based on an assumption that observations are random samples from a normal distribution, known apart from one or both parameters, may not be seriously astray even if the normality assumption is incorrect. But this is not always true.

We sometimes want to make inferences that have nothing to do with parameters, or we may have data in a form that makes, say, normal theory tests inappropriate; we may not have precise measurement data, but only the rank order of observations. For example, although it is often reasonable to assume examination marks are approximately normally distributed, these marks may not be published. We may only know the numbers of candidates in banded and ordered grades designated Grade A, Grade B, Grade C, ..., or Level I, Level II, etc. In Example 1.1 we consider a situation where we know only total numbers of items and the proportions with a certain characteristic.

Even when we have precise measurements it may be obvious that we cannot assume a normal distribution. We may be able to say little except perhaps that the distribution is skew, or symmetric, or has some other characteristic.

Appropriate methods of inference in these situations are described as **nonparametric**, or sometimes more aptly, as **distribution-free**.

Many writers regard 'distribution-free' and 'nonparametric' as synonyms, a view that ignores subtle distinctions that need not worry us here. Some tests that are generally classed as nonparametric or distribution-free do indeed involve parameters and distributions and the 'distribution-free' or 'nonparametric' tag simply means they can be applied to samples that come from populations having any one of a wide class of distributions. In general, these methods are applicable to estimation or hypothesis-testing problems when the population distributions need only be specified in broad terms, e.g. as being continuous, symmetric, identical, differing only in median or mean; they need not belong to specific families such as the normal, uniform, exponential, etc. Logically, the term distribution-free may then be more appropriate than nonparametric, but the latter term is well established in popular usage.

1.1.2 Why do we need nonparametric methods?

A parametric test may depend crucially for its validity on an assumption that we are sampling randomly from a distribution belonging to a particular family. If there is doubt a nonparametric test that is valid under weaker assumptions is preferable. Nonparametric methods are invaluable – indeed they are usually the only methods available when we have data that specify just order or ranks or proportions and not precise observational values.

It must be stressed that weaker assumptions do not mean (as research workers sometimes misguidedly think) that nonparametric methods are assumption-free. What can be deduced depends on what assumptions can validly be made; an example demonstrates this.

Example 1.1

A manufacturer mass-produces an item that has a nominal weight of 1000 g and gives a guarantee that in large batches not more than 2.5% will weigh less than 990 g. The plant is highly automated and to check that the two machines being used are producing goods of acceptable quality the manufacturer takes samples of 500 at regular intervals from the production run for each machine. These are put through a quick-operating electronically controlled checker that rejects all items from the 500 that weigh less than 990 g. This provides the only check that the requirement is being met that not more than 2.5% are below 990 g. This is a typical observation for a quality control programme.

To give reasonable protection the machines may be adjusted to produce not more than 2.25% underweight items (i.e. below 990 g) when operating properly. If underweight items are produced at random and the target of 2.25% is maintained for a large batch, then the number underweight in samples of 500 will have a binomial distribution with $n = 500$ and $p = 0.0225$. Standard quality control methods use such information to indicate if a batch is reasonably likely to meet the guarantee criterion; such test procedures are parametric, based on the binomial distribution.

But the manufacturer may ask if other deductions can be made from the test information. For example, do the numbers underweight throw light on the underlying distribution of weights? For example, can we use the observed numbers underweight in samples of 500 from each of the two machines to test whether the mean weight of items produced by each machine are equal, or have a specific value? We cannot do this without further assumptions about the distributions of weights for each machine. This is immediately apparent because the proportion weighing less than 990 g will be 2.25% for an infinity of possible distributions. For example, if the weights are normally distributed with a mean of 1000 g and a standard deviation of 5 g then the long-run proportion below 990 g can be shown to be 2.25% (more exactly 2.28%, but for simplicity we ignore this rounding difference). Also, if the weights were distributed normally with a mean of 1005 g and standard deviation 7.5 g, the proportion below 990 g is again 2.25%, as it also would be if the weights had a uniform (rectangular) distribution over the interval (989.55, 1009.55). We could find not only an infinity of other normal or uniform distributions, but also gamma and other distributions which gave the same proportion below 900 g. For all of these the binomial distribution with $n = 500$ and $p = 0.0225$ is relevant to the distribution of underweight items in our samples. Tests based on the binomial distribution cannot therefore tell us on their own if the two machines are producing items with the same mean weight. However, if we now make an additional assumption that the output from the two machines have **identical distributions apart perhaps from a shift in median if something has gone wrong with one**, we may use our binomial-type information to test the hypothesis that the medians are identical, against the alternative that the two population medians differ. This is a nonparametric test, but it would not be very efficient. If, from past experience, the manufacturer could say that the weights always has a normal distribution with known standard deviation, say 6, but the mean was liable to shift if things went wrong, then knowing the number of items with weight below 990 g in a sample of 500 enables one to test a hypothesis that the mean is 1002 (the target value to give 2.25% defectives with standard deviation 6) against the alternative that the mean had some other value. This would be a parametric test about a parameter μ, the mean of a normal distribution with standard deviation 6. Again, it would not be a very efficient test, but the best we could do without more detailed information about weights of the items sampled. We say more about hypothesis testing in Section 1.2.

Means and medians (see Section A1.2) are common measures of location. The most common problems with measurement data concern **location**. Is it reasonable to suppose a sample comes from a population with a certain specified mean or median? Can we reasonably assert that two samples come from populations whose means differ by at least 10 units? Given a sample, what is an appropriate estimate of the population mean or median? How good an estimate is it?

Some nonparametric methods require only minimal information. To test whether we may reasonably assert that a sample might be drawn from a distribution with pre-specified median θ, say, we need only know how many sample values are greater than θ and how many are less than θ. If it were difficult or expensive to get exact observations, but easy to determine numbers above or below θ, this nonparametric approach may be very cost-effective.

Example 1.2

I have on my bookshelf 114 books on statistics. I take a random sample of 12 of these (i.e. the 12 are selected by a method that gives each of the 114 books an equal chance of selection – see Section A3). I want to test the hypothesis that the median number of pages per volume is 220.

I check in each of the 12 volumes selected whether or not the number of pages exceeds 220. In 9 it does, in the remaining 3 it does not. I record this as 9 pluses (+), or excesses over the hypothetical value, and 3 minuses (−).

A population median of 220 would imply half of the 114 books on my shelf have 220 or less pages, half that or more. This follows from the definition of a median (Section A1.2). Thus, if the median were 220, when we select a random sample it is (for practical purposes) equally likely that each book selected will have more than 220 pages (scored as +) or less (scored as −). A complication we discuss in Section 2.1.1 occurs if a sampled book has exactly 220 pages; this did not happen in my sample.

By associating a plus with a 'head' and a minus with a 'tail' we have a physical analogue with coin tossing; if the median is really 220, then the result 9 plus and 3 minus signs is physically equivalent to 9 heads and 3 tails when a fair coin is tossed 12 times. We show in Section 1.2 that this evidence does not justify rejection of the hypothesis that the population median is 220.

Non-rejection of a hypothesis in this sense does not **prove** it is true; it only means that currently we have insufficient evidence to reject it. We do not reject, because 9 heads and 3 tails is in a set of reasonably likely results when we toss a true coin. Had we got 12 plus and no minus signs, or vice versa (equivalent to 12 heads or 12 tails) we could reasonably reject the hypothesis that the median is 220. Indeed, the probability of getting one or other of these extremes in a sample of 12 is only 1/2048, so such a result in just one experiment means either we have observed an event with odds heavily stacked against it, or our hypothesis of a fair coin (or that the median is 220) is not correct. The latter seems more plausible. For those who are not already familiar with these ideas we formalize them in Section 1.2. The relevant test is called the **sign test**.

1.1.3 An historical note

It is fashionable to claim that nonparametric methods were first used when J. Arbuthnot (1710) found that in each year from 1629 to 1710 the number of males christened in London exceeded the number of females. He regarded this as strong evidence that the probabilities of any birth being male or female were not exactly equal, a discrepancy Arbuthnot attributed to 'divine providence'. A sign test is appropriate for his data.

Spearman (1904) proposed a rank correlation coefficient that bears his name, but a serious study of nonparametric methods for statistical inference began some fifty years ago in an era when applied statistical methods were dominated by grossly oversimplified mathematical models of real-world situations chosen partly because they led to not too demanding computational procedures: the inaptly named normal distribution was the key to analytical methods for continuous data; the binomial and Poisson distributions to methods for discrete data. These distributions still are – and always will be – important, but they are not all-embracing.

Research into nonparametric and distribution-free methods was stimulated firstly by attempts to show that even if assumptions of normality often stretched credulity, then at least in some cases making those assumptions would not greatly alter valid inferences. This was the stimulus of work by R. A. Fisher, E. J. G. Pitman and B. L. Welch on randomization or permutation tests; tests which at that time (the 1930s) were computationally too demanding for general use.

At about the same time there was a growing realization that observational data that were not numerically precise but consisted of preferences or rankings could be used to make inferences in a way that required little computation effort.

A few years later F. Wilcoxon and others realized that even if we had precise numerical data we sometimes lost little useful information by replacing them by their rank order and basing our analysis on computationally simple procedures using these ranks. Indeed, if an assumption of normality were not justified, analyses based on ranks were sometimes the most efficient available. This heralded an era when nonparametric methods developed as practical tools for use either when data were by nature simply ordinal (ranks or preferences) as distinct from precise measurements (interval or scalar); or as a reasonably efficient method that reduced computation even when full numerical data were available, but could easily by replaced by ranks. Used in this way there were still many limitations: simple hypothesis testing was usually easy; interval estimation much more difficult.

Ever-increasing calculating power of modern computers has revolutionized our approach to data analysis and statistical inference (see e.g. Durbin, 1987). Pious hope that data fit a restricted mathematical model with few parameters and emphasis on simplifying concepts such as linearity have been replaced by the use of robust methods and by exploratory data analysis in which we investigate different potential models; areas where nonparametric methods have a central role.

They may also be applied to counts, these often recorded as numbers of items in various categories; e.g. numbers of examination candidates obtaining Grade A, Grade B, Grade C passes. Here the categories are **ordinal**; Grade A is better than Grade B; Grade B is better than Grade C; and so on. Categories that cannot be ordered by the inequalities greater than or less than are called **nominal**; e.g. people may be classified as single, married, widowed or divorced. For data in these forms nearly all analyses are by nature nonparametric.

A disadvantage of nonparametric methods in the pre-computer era was that simplicity only applied to basic procedures and nonparametric methods lacked the flexibility of much linear model and least squares theory that are cornerstones of normal distribution parametric inference. The advent of computers has revolutionized this aspect of using nonparametric methods, for many advanced and flexible methods are tedious only in that they require repeated application of simple calculations – a task for which computers are admirably suited and easily programmed.

The dramatic post-war development of nonparametric methods is described by Noether (1984). Some idea of the volume of literature is given in the nonparametric bibliography compiled by Singer (1979). It considers work relevant to applications in just one subject – psychology. Work continues at an increasing pace.

1.2 HYPOTHESIS TESTS

We assume a basic familiarity with hypothesis testing like that implicit in the use of t-tests, but we summarize a few fundamentals and illustrate application to a simple nonparametric situation.

In Example 1.2 we wanted to test acceptability of a hypothesis that the median number of pages in all 114 books was 220. This implies the median is some other number if that hypothesis is not true, so our hypothesis that it equals 220 is something of a cockshy. It may have been suggested by past experience in assessing book lengths, or have been asserted with confidence by somebody else. We call this a **null hypothesis**, writing it H_0. If θ denotes the population median we often use the shorthand notation

$$H_0: \theta = 220$$

Our alternative hypotheses, collectively labelled H_1, are written

$$H_1: \theta \neq 220$$

We speak of testing H_0 against the **two-sided** (greater or less than) alternatives H_1.

The sign test in Example 1.2 involved a known distribution of signs if the null hypothesis H_0 were true, namely the binomial with $n = 12$ and $p = \frac{1}{2}$. The probabilities of each number of successes (here represented by plus signs) is tabulated; see, e.g. Neave (1981, p. 6). They are given (to three decimal places) in Table 1.1.

From Table 1.1 we see that if H_0 is true, 6 plus (hence also 6 minus) has maximum probability, and the probabilities fall off

Table 1.1 Binomial probabilities, P, for r plus signs, $n = 12$, $p = \frac{1}{2}$

r	0	1	2	3	4	5	6	7	8	9	10	11	12
p	.000	.003	.016	.054	.121	.193	.226	.193	.121	.054	.016	.003	.000

symmetrically to be negligibly small at 0 or 12 plus signs (the exact values in each case then being about 0.000 24). As already implied, these extreme results provide strong evidence against H_0. In particular, 12 plus signs suggest a higher median value than 220; 12 minus signs a lower value. In formal hypothesis testing we divide the outcomes in Table 1.1 into two sub-sets. The first contains the most likely outcomes under the null hypothesis with a preassigned associated total probability which is conventionally at least one of the values 0.90, 0.95, 0.99 or 0.999. Of these values the most commonly used is 0.95. It is easily verified from Table 1.1 that the set from 3 to 9 has associated total probability 0.962; excluding any of these outcomes (which in this symmetric case can only be done fairly on a pairwise basis) reduces the probability below 0.95. The remaining outcomes (i.e. 0, 1, 2, 10, 11, 12) form a set with associated total probability 0.038 ($= 1 - 0.962$). We call the latter set a **critical region** of **nominal size** 0.05 and of **actual size** 0.038. Ideally we like critical regions to have the same actual and nominal sizes; in practice we can often (as in this case) only make them so by artificial dodges that generally do more harm than good. The hypothesis-testing rule is that we reject the null hypothesis H_0 at a (nominal) 0.05 (or 1 in 20, or 5%) significance level if (and only if) our result falls in the critical region. One may, if one wishes, rephrase this in terms of an (actual) 0.038 (or 3.8%) significance level, but the former terminology, using nominal rather than actual, is more usual for practical reasons. Commonly used nominal levels are 0.05, 0.01, 0.001 (equivalently 5%, 1%, 0.1%). Occasionally the level 0.10 (10%) is also used.

We have framed the argument in terms of probabilities, but for convenience throughout the book we shall henceforth express significance levels as percentages. Corresponding to a critical region of size α (a probability) we shall speak of a $100\alpha\%$ significance level.

In our example the set of outcomes (0, 1, 11, 12) has associated probability of 0.006 and is a critical region of actual size 0.006, or nominal size 0.01, for testing H_0 against the two-sided alternative H_1. The complementary set of all outcomes not in the critical region forms a **non-rejection** (sometimes called an **acceptance**) **region** for the null hypothesis. A test of the type just described is often called a **two-tail** test, for the critical region is formed from the two tails of the distribution appropriate to the null hypothesis. Two-tail tests are associated with two-sided H_1. In this particular example, because of symmetry, we take the same number of points in each tail. More generally, if we are seekings a critical region of nominal size α for a two-tail test we select points from each tail with associated probability as close as possible to but not exceeding $\frac{1}{2}\alpha$ from each tail. Arguments can be made for alternative procedures but the differences are usually small. By selecting a critical region with actual size less than or equal to our nominal α we take a conservative attitude. Some writers choose the actual size to be as close as possible to α. Thus if $\alpha = 0.05$ and possible actual sizes are 0.032 and 0.061, they choose the actual size 0.061. This enhances the probability of rejection at the given nominal level. Ideally one would prefer always to quote actual levels. A practical reason why this is often not possible is that most published tables refer only to nominal levels; without a versatile computer program, it may be difficult to evaluate actual levels. Further, we often use approximate tests where the approximation may introduce a discrepancy as great as that between actual and nominal levels. The moral is that we should not sanctify commonly used levels such as 5%, but regard them more as useful quantitative guidelines. Fifty years ago such an attitude would have been deemed statistical heresy.

1.2.1 One-tail tests

There are many situations when we are only interested in testing a hypothesis of the form

$$H_0: \theta = \theta_0$$

against a one-sided alternative of the form

$$H_1: \theta > \theta_0$$

We use what is called a **one-tail** test.

This would be appropriate in a situation modified from that in Example 1.2 if we knew from past experience (or by inspecting all 114 books) that the median length of all my books was indeed 220 pages but I am now given a further 50 books by a friend who no longer wanted them; because each appears rather long, I suspect the median for my enlarged collection is increased. (I may, for example, have observed that the shortest of the new books has 212 pages). In these circumstances, while the median length of all books may well exceed 220 pages, it is almost inconceivable that it will be reduced by the extra books.

Clearly a surplus of positive signs in a sample from the combined collection would indicate H_1 might be true. We now choose a critical region from Table 1.1 from the top tail only, with an associated probability as close as possible but not exceeding 0.05 if we want to test at the 5% significance level. As it happens, we still only include the points 10, 11, 12, but the associated probability is reduced to 0.019, well below the nominal level. If we are prepared to accept significance at the 10% level (and this may be unwise for reasons that become clearer in Section 1.2.2) our critical region would contain the points 9,

10, 11, 12, with actual size 0.073. This represents a true significance of 7.3%, tighter than the nominal 10%. In Example 1.2 we had 9 plus signs, so in a one-sided test using a nominal 10% (actual 7.3%) level we would have rejected the null hypothesis. Of course in a given situation either a one- or two-tail test is appropriate – not both. Which is appropriate depends on the question being asked, and the relevance of the data and the experimental set-up, to answering that question. See Lehmann (1975, p. 24) for a sound discussion of this matter. We consider this question further in comment 2 on Example 2.1 in Section 2.1.1. A fairly obvious modification is needed if our alternative hypothesis is $H_1 : \theta < \theta_0$.

When we express our hypotheses $H_0 : \theta = \theta_0$; $H_1 : \theta > \theta_0$, what happens if $\theta < \theta_0$? To allow for this possibility we should write $H_0 : \theta \leq \theta_0$. The conduct of the test is not altered. The probability of getting a result in the upper tail is greater when $\theta = \theta_0$ than it is if $\theta < \theta_0$, so we are less likely to erroneously accept H_1 when $\theta < \theta_0$ than would be the case if $\theta = \theta_0$.

1.2.2 Kinds of error

In hypotheses tests two kinds of error are possible. We may reject the null hypothesis, H_0, when it is true. This happens if our test result falls in the critical region (the region for rejection) when the null hypothesis is true. The probability of this is given by the size of the critical region. By choosing a critical region of nominal size 0.05 there is at most 1 chance in 20 of rejecting a true null hypothesis; this is why we sometimes speak of significance at the 1-in-20 level as equivalent to the 0.05 probability level or 5% level. When we wrongly reject a null hypothesis we make an **error of the first kind**.

We make **an error of the second kind** if we do not reject the null hypothesis when it is false. This occurs if our result falls in the non-rejection, or acceptance, region when H_0 is false. We would prefer that our result fell in the critical region when H_0 is false, i.e. when H_1 is true. When H_1 is of a form $H_1 : \theta \neq \theta_0$ or $H_1 : \theta > \theta_0$ or $H_1 : \theta < \theta_0$, there is an infinity of possible values for the true θ and the probability of rejecting H_0 when it is false usually depends strongly on which particular θ among the many included in H_1 is the true value.

Specifically, in tossing a coin 12 times, we are clearly more likely to get 10, 11, or 12 heads if, at each toss, the probability of heads, $p = 0.9$, than would be the case if it were $p = 0.55$, although in each case the hypothesis $H_0 : p = 0.5$ is false. In the extreme case when a coin is a double-header we are certain to get 12 heads and therefore to reject H_0 at the 5% (even at the 0.1%) significance level with either a one- or two-tail test.

The probability of getting a result in the critical region when a specified single value in H_1 is the true θ value is called the **power** of the test for H_0 against that particular alternative. The power depends on:

1. the significance level of the test
2. the true value of the parameter or other measure we are testing
3. the size of the sample
4. the particular test we are using.

We give no detailed theory of power, but there are a number of good practical indications of how a test is likely to perform in broad terms. Generally speaking, the more relevant information we incorporate validly into a test, the higher the power. For example, the sign test for the median in Example 1.2 is usually less powerful than an appropriate test based on the actual numbers of pages in each volume included in the sample. If we could validly assume that numbers of pages in statistics books are approximately normally distributed, a t-test using the same sample would be more powerful than any other test; i.e. would be most likely to detect a given departure from H_0.

Theoretical results indicate that if we double the sample size for the sign test it will often have more power than the t-test applied to the smaller sample even if we can validly assume normality. If valid, the t-test would do better than the sign test if both were applied to the larger sample.

For cases where we have actual measurements but cannot assume normality we discuss other tests in Chapter 2 that are generally more powerful than a sign test based on the same number of observations.

1.2.3 Pitman efficiency

In Section 1.2.2 we listed factors that influence the power of a test. Most 'intuitively reasonable' tests have good power to detect a true alternative that is far removed from the null hypothesis. The sign test soon detects a double-headed coin! In practice we want as much power as possible for detecting alternatives close to H_0. Pitman (1948), in a series of unpublished lecture notes, introduced the concept of **asymptotic relative efficiency** for comparing two tests. It is based on the idea of **relative efficiency** of any two tests, T_1 and T_2. If α is the probability of an error of the first kind and β is the probability of an error of the second kind (the power is $1 - \beta$), then the efficiency of T_2 relative to T_1 is the ratio n_1/n_2 of the sample sizes needed to attain the same power for the two tests with the chosen α, β. In practice we usually fix α; then β depends on the particular alternative as well as the sample sizes. Fresh calculations of relative efficiency are required for each particular value of the parameter (or parameters) of interest in H_1 and for each choice of α, β.

Pitman considered sequences of tests T_1, T_2 in which we first fix α, but allow the alternative in the set H_1 to vary in such a way that β remains constant as the sample size n_1 increases. For each n_1 we determine n_2 such that T_2 has the same β for the particular alternative considered. It is intuitively obvious that as we take larger samples we will increase the power for alternatives close to H_0, so that when the samples are very large we study the behaviour of the efficiency, n_1/n_2, for steadily improving tests for finding small departures from H_0. Pitman found that in these sequences of tests, under very general conditions, n_1/n_2 tended to a limit as $n_1 \to \infty$, and more importantly, that this limit was the same for all choices of α, β. The limit is the asymptotic relative efficiency (ARE) of the two tests. An alternative definition of asymptotic relative efficiency has been proposed by Bahadur (1967) but is less widely used. For clarity and brevity we refer in this book to Pitman's asymptotic relative efficiency of two tests simply as the **Pitman efficiency**.

What use is a result for large samples when in practice we often deal with small ones? The practical usefulness lies in the fact that when comparing many tests the small-sample relative efficiency is often close to, or even better than, the Pitman efficiency.

The Pitman efficiency of the sign test relative to the t-test when the latter is appropriate is $2/\pi = 0.64$. Lehmann (1975, p. 173) shows that for samples of size 10 and a wide range of true values of the median θ relative to θ_0 with α fixed, the relative efficiency exceeds 0.7, while for samples of 20 it is nearer to, but still slightly above, 0.64. In that case the Pitman efficiency gives a pessimistic picture of the performance of the sign test at realistic sample sizes.

It is well known that when the t-test is relevant and valid it is the most powerful of all tests for any H_0 against any alternative H_1. There are situations when it is not appropriate; then the sign test may have higher efficiency. Indeed, if our sample comes from a distribution called the double exponential which has longer tails than the normal, the Pitman efficiency is 2. That is, we would do as well (at least for large samples) with a sample half the size using a sign test as we would with a t-test on the larger sample. In practice, we often have little idea, except in vague terms, of the distribution, or population, from which we sample, but it is broadly true that if we suspect our sample comes from a long-tail distribution (i.e. with tails longer than the normal) we may do better for location tests or estimates (i.e. those concerning means or medians) with nonparametric tests.

1.3 ESTIMATION

1.3.1 Confidence intervals

Although the estimation problem is commonly formulated in different terms, one way of specifying a $100(1 - \alpha)\%$ confidence interval for a location parameter θ (see Section A1.2) is to define it as a set of all values, θ, which would be accepted using a hypothesis test at significance level $100\alpha\%$.

For the sign test for a median using a sample of 12 and a two-tail test at the 5% level we saw that we accepted H_0 if we got between 3 and 9 plus signs inclusive. If we know the actual sample values we can establish confidence intervals immediately. For the sample of books considered in Example 1.2 these values (arranged for convenience in ascending order) were (see data set in Section A7.1):

$$126, 142, 156, 228, 245, 246, 370, 419, 433, 454, 478, 503$$

It is immediately clear that we have between 3 and 9 plus signs for any values of θ greater than 156 and less than 454. The open interval $(156, 454)$ is thus a nominal 95% (actual $100(1 - 0.038)\% = 96.2\%$) confidence interval for θ, the population median.

We need not assume our sample comes from a population with a symmetric distribution to apply the sign test, but if we do then the population mean and median coincide. In Example 1.2 if we assume symmetry and also that numbers of pages are approximately normally distributed, it is easily verified that a 95% confidence interval based on the t-distribution and the above sample is $(227.4, 405.9)$; see Section A4.1. This is a shorter interval than that based on the sign test, reflecting our use of more detailed information, but we may wish to consider the assumption of normality before accepting its validity. Note that a t-test (unlike the sign test) would reject the hypothesis $\theta = 220$ at the 5% significance level, since 220 lies outside the 95% confidence interval.

The more usual interpretation of the $100(1 - \alpha)\%$ confidence interval is that of an interval having the property that if we consistently formed such intervals for repeated samples then $100(1 - \alpha)\%$ of such intervals would contain the true unknown θ.

The numbers of pages for all 114 books are given in Section A7.1; we find the true median in Exercise 1.2.

1.3.2 Tolerance intervals

Another useful concept is that of a **tolerance interval**. These are intervals with properties of the form: given p_1, α there is a probability of at least $1 - \alpha$ that the interval includes $p_1\%$ of the population. Given a sample of n observations $x_1, x_2, \ldots, x_n$ from a continuous distribution, the tolerance interval concept is used to answer such questions as: how large should n be so that the probability is at least 0.95 that 90% of the population lies between the smallest and largest sample value? Or, how large a sample is needed to ensure that with probability 0.8 at least 75% of the population exceeds the second smallest sample value? Procedures for problems about tolerance intervals and limits are discussed by Conover (1980, Section 3.3).

1.4 SAMPLES AND POPULATIONS

The random sample of 12 books from a population of 114 statistics books on the shelf in my study considered in Example 1.2 is a sample from a finite population. However, we might be prepared to assume that inferences about numbers of pages in the books on my shelves hold good for a much wider group of statistics books – perhaps all those published in the English language in the last 50 years. This would be perfectly valid if my own set of 114 books were itself a random sample of all such books, but I do not buy books at random. However, my interests in the subject are fairly wide, so it is likely that my books are a pretty good cross-section of those published on the subject and so are probably not too unlike a typical random sample.

Many people would therefore be prepared to believe that hypotheses about numbers of pages applicable to my books apply to a more general class of books on statistics. Nevertheless, caution is needed about extrapolating to larger populations than that involved in an experiment. For example, if all my books had been on applications of statistics I suspect the median number of pages would not be typical of that of all statistics texts because there are many highly specialized books on the theory that are often, by their nature, very short (and a few very long and boring). These would not be represented in anything like the correct proportion in a library devoted mainly to applied statistics.

It is doubtful whether findings about median lengths of statistics books could be extrapolated to a wider class of books such as general fiction or reference works.

We often face similar situations. In comparing the effect of several diets on growths of pigs of a particular breed we may use in an experiment a sample of 50 pigs from some very select group but hope our results will apply if the feeds were given under similar conditions to all pigs of that breed.

The abstract notion of sampling from an infinite population (implicit in normal theory) is one that often works well in practice but is never completely true.

At the other extreme there are situations where we are essentially restricted to experiments that involve not a sample from any larger population but where our sample units are essentially the whole population. This often happens at the early stages of testing a new drug. If it is to be tested for treatment of a rare disease there may be just, say, eight patients available. If the patients all suffer the disease with approximately equal severity a common way to test the drug is to allocate, say, four of these patients to receive the drug. The remaining four are untreated (or perhaps treated with a standard drug already in common use). If the four patients who receive the new drug are selected at random (from the eight) and the drug has no effect (or is no better than one currently in use) it is unlikely that a later examination would indicate that the four patients receiving the new drug had responded better than any of the others. This is a possibility, but it has a low probability which we can calculate.

1.4.1 Permutation tests

Example 1.3

Suppose, as suggested above, we have eight patients and four are selected at random to receive the new drug. After three weeks all eight patients are examined by a skilled consultant. For some diseases there will be one or more criteria (blood pressure, sugar content of urine, etc.) that give an indication of the patients' condition. In other cases the doctor may only be able to 'rank' the patients in order from less severe (ranked 1) to most severe (ranked 8). In this latter situation if there is really no beneficial effect of the new treatment, what is the probability that the patients ranked 1, 2, 3 and 4 after treatment are those who received the new treatment?

Selection of four patients at random means any four are equally likely to be chosen for the new treatment. If there really is no effect one would expect some of those chosen to end up with low rankings, and others with high rankings.

From a group of eight patients numbered 1 to 8 there are 70 different ways of selecting sets of four. This is an application of the result in Section A2 for calculating the number of ways of selecting r objects from n. We give the 70 selections in Table 1.2. Ignore for the moment the numbers in brackets after each selection.

If the new drug is ineffective the set of ranks associated with the four patients receiving it is equally likely to be any of the 70 quadruplets listed in Table 1.2. Thus if there is no treatment effect there is just 1 chance in 70 that the four showing greatest improvement (ranked 1, 2, 3, 4 in order of condition after

Table 1.2 Possible selections of four numbers from eight labelled 1 to 8

1,2,3,4 (10)	1,2,3,5 (11)	1,2,3,6 (12)	1,2,3,7 (13)	1,2,3,8 (14)
1,2,4,5 (12)	1,2,4,6 (13)	1,2,4,7 (14)	1,2,4,8 (15)	1,2,5,6 (14)
1,2,5,7 (15)	1,2,5,8 (16)	1,2,6,7 (16)	1,2,6,8 (17)	1,2,7,8 (18)
1,3,4,5 (13)	1,3,4,6 (14)	1,3,4,7 (15)	1,3,4,8 (16)	1,3,5,6 (15)
1,3,5,7 (16)	1,3,5,8 (17)	1,3,6,7 (17)	1,3,6,8 (18)	1,3,7,8 (19)
1,4,5,6 (16)	1,4,5,7 (17)	1,4,5,8 (18)	1,4,6,7 (18)	1,4,6,8 (19)
1,4,7,8 (20)	1,5,6,7 (19)	1,5,6,8 (20)	1,5,7,8 (21)	1,6,7,8 (22)
2,3,4,5 (14)	2,3,4,6 (15)	2,3,4,7 (16)	2,3,4,8 (17)	2,3,5,6 (16)
2,3,5,7 (17)	2,3,5,8 (18)	2,3,6,7 (18)	2,3,6,8 (19)	2,3,7,8 (20)
2,4,5,6 (17)	2,4,5,7 (18)	2,4,5,8 (19)	2,4,6,7 (19)	2,4,6,8 (20)
2,4,7,8 (21)	2,5,6,7 (20)	2,5,6,8 (21)	2,5,7,8 (22)	2,6,7,8 (23)
3,4,5,6 (18)	3,4,5,7 (19)	3,4,5,8 (20)	3,4,6,7 (20)	3,4,6,8 (21)
3,4,7,8 (22)	3,5,6,7 (21)	3,5,6,8 (22)	3,5,7,8 (23)	3,6,7,8 (24)
4,5,6,7 (22)	4,5,6,8 (23)	4,5,7,8 (24)	4,6,7,8 (25)	5,6,7,8 (26)

treatment) happen to be those allocated to the new treatment. It seems more plausible that such an outcome reflects a beneficial effect of treatment.

Let us assemble these ideas into an hypothesis-testing framework. With a comparison between two groups of four, each group treated with a different drug, we have what is called a two independent sample experiment. This we discuss in detail in Chapter 5. Under the null hypothesis of no difference in the treatments we reiterate that we are likely to find a fair mixture of ranks in the final grading for those receiving the treatment, some patients doing better with the new drug, others less well. We have seen that the most favourable outcome for the new drug – that the patients receiving it are ranked 1, 2, 3, 4 in order of later fitness – has only 1 chance in 70 of happening if there is no real effect. If we get this result, using a one-tail test of the null hypothesis

$$H_0: \text{New treatment is non-effective}$$

against the alternative

$$H_1: \text{New treatment is beneficial}$$

we have significance at an actual 1 in 70 level since clearly the result is in an appropriate critical region of this size for a one-tail test.

What if the patients receiving the new treatment were ranked 1, 2, 3, 5? Again we intuitively feel this is evidence favouring the new treatment.

In practice we set about deciding what set of ranks indicate rejection of H_0 by looking at some function of the ranks that will have a low value if all ranks are low, a medium to high value otherwise. An obvious choice is the sum of the ranks. If we sum the ranks for every quadruplet in Table 1.2 and count how many times each sum occurs we can work out the probability of getting each sum.

In Table 1.2 the numbers in brackets after each quadruplet are the sum of the ranks in that quadruplet, e.g. for 1, 2, 3, 7 the sum is $1 + 2 + 3 + 7 = 13$. The lowest sum (10) is for the quadruplet 1, 2, 3, 4 and the highest (26) for the quadruplet 5, 6, 7, 8. Table 1.3 gives the number of quadruplets having each given sum.

Intuitively, low sums indicate the drug is beneficial and high sums that it is deleterious. For the hypothesis test outlined above, a one-tail critical region of nominal size 0.05 is given by rank sums 10 and 11 (actual size $2/70 = 0.029$).

Table 1.3 Sums of ranks of four items from eight

Sum of ranks	10	11	12	13	14	15	16	17	18
No. of occurrences	1	1	2	3	5	5	7	7	8

Sum of ranks	19	20	21	22	23	24	25	26
No. of occurrences	7	7	5	5	4	2	1	1

The sums 10, 11, 12 would provide a critical region of actual size $4/70 = 0.057$, which would be accepted as a good approximation to 0.05 by those who work with nearest actual size rather than an actual size not exceeding 0.05. If our alternative hypothesis is that the new drug may be beneficial or deleterious a two-tail test would be appropriate (see Exercise 1.4).

Tests based on permutations of ranks (and sometimes permutations of actual observations) and functions like rank sums play a large part in nonparametric methods and are called **permutation** or **randomization** tests. In this particular example the test is applied to the population of eight patients in the trial which were divided into two samples. There is no obvious larger population from whom the eight are a random sample to which we can validly make inferences. Efficacy of a drug may depend on a number of factors such as severity of disease, treatment being administered sufficiently early, the age and sex of patients, etc., all or none of which may be adequately reflected in a small selection of 'available patients'. An encouraging result with this small group would suggest further experiments were desirable. We look in more detail at permutation tests for single samples in Section 3.3, Example 3.6, and for two samples, using ranks, in Section 5.1.2.

1.5 FURTHER READING

This book is a guide to practical application of nonparametric methods. We give simple explanations of the rationale behind procedures, but omit detailed theoretical explanations (often involving complicated mathematics). There are a number of excellent texts that pursue these matters. The book by Conover (1980) gives more background for many of the procedures described here and precise information on when each is applicable, and is highly recommended for more detailed study. Lehmann (1975) discusses the rationale and theory carefully, using simple illustrative numerical examples without advanced mathematics. His book repays careful reading for those who want to pursue the logic of the subject in more depth. Randles and Wolfe (1979) and Maritz (1981) are excellent books covering the theory at a more advanced mathematical level. Daniel (1978) and Hollander and Wolfe (1973) are standard works on applied nonparametric methods, the latter at the more advanced mathematical level. Marascuilo and McSweeney (1977) and Leach (1979) concentrate on applications in the social sciences; as does the classic, but now somewhat dated yet still very readable text by Siegel (1956). Bradley (1968) is another pioneer text.

1.6 COMPUTERS AND NONPARAMETRIC METHODS

The advent of computers has increased the scope of nonparametric methods. It will be apparent that many of the methods for confidence limits developed in later chapters will often only be practicable if a computer is available. Indeed, they are ideally suited to computing, involving ordering of data and repetitions of simple yet tedious calculations, both tasks at which computers excel, and which are easily programmed.

Many standard computer packages, including recent versions of SPSS and Minitab, both for mainframe and PCs, include some nonparametric programs, and many techniques described in this book are not difficult to program in standard languages such as Basic, Fortran or Pascal. More specialist methods like those involving log-linear models (Section 8.3) are included in GLIM.

The situation regarding available packages and their updating is constantly changing. Relevant documentation should indicate what each package does. Users of a particular package may find it informative to test them using examples from this book to ensure that they can interpret relevant output. In some cases the output will inevitably be in different form to that given here; being either more or less extensive. If so, it is important to understand the reason for any differences. For example, output may give actual rather than nominal significance levels, or these and confidence intervals may be adjusted to bring actual and nominal levels closer than the basic distribution theory allows. These are problems that do not arise with normal theory tests like the t-test where nominal and actual levels coincide. A further complication is that sometimes alternative statistics (usually transformations of one another) are used by different authors. For example, in Section 7.1.2 we recommend basing tests for correlation on a function of Kendall's tau, but tests may be based on the statistic tau itself.

In many sophisticated applications of nonparametric methods such as those outlined in Chapters 9 and 10 a computer is virtually essential for implementation.

EXERCISES

1.1 Use the numbers of pages given on p. 13 for my sample of 12 books to perform a sign test of the hypothesis $H_0:\theta = 460$ against $H_1:\theta \neq 460$, testing at a nominal 5% significance level. Reconcile your finding with the confidence interval given on p. 13. Without actually performing a t-test use information given on p. 13 to state whether or not you would reject $H_0:\theta = 460$ if you had used a t-test.

1.2 Determine the median number of pages for all 114 books given in Section A7.1. Does the true median lie within the confidence intervals given in Section 1.3.1?

1.3 Use the sums of ranks given in brackets after each group in Table 1.2 to verify the correctness of the entries in Table 1.3.

1.4 Determine a critical region based on sums of ranks of nominal size 0.05 for testing the hypothesis H_0: drug has no beneficial effect, against H_1: drug has an effect (beneficial or deleterious) in the context of Example 1.3.

1.5 An archaeologist numbers some articles 1 to 11 in the order he discovers them. He selects at random (without replacement) a sample of three of them. What is the probability that the sum of the numbers on the items he selects is less than or equal to 8? (It is not necessary to list all combinations of 3 items from 11 to answer this question.)

1.6 In section 1.3 we associated a confidence interval with a two-tail test. As well as such two-sided confidence intervals, one may define a one-sided confidence interval composed of all parameter values that would not be rejected in a one-tail test. Follow through such an argument based on the sign test criterion for the 12-book sample values given in Section 1.3.1 relevant to a one-sided alternative $H_1 : \theta > \theta_0$.

2

Location estimates for single samples

2.1 THE SIGN TEST

In Chapters 2 to 7 we consider data that are either measurements or ranks specifying order of magnitude or preference. The latter are called **ordinal** data. Chapters 2 and 3 are devoted to single samples. Most practical problems involve comparison of, or studying relations between, several samples, but many basic nonparametric notions are applicable also to a single sample and the logic behind them is easily explained in this basic situation.

We introduced the sign test informally in Example 1.2. We look more closely at assumptions and at difficulties or complications that may occur and extend the concept to related problems.

We develop arguments largely by specific examples. In most of these we discuss points under these headings:

The problem
Formulation and assumptions
Procedure
Conclusion
Comments

We separate a particular example from general text by a line space after the comments.

At the end of each chapter we list, before the exercises, some common fields of application of the techniques; these lists are indicative rather than exhaustive.

2.1.1 The effect of sample size

For the sign test – and indeed for practically all the methods in this book – a larger sample increases the power of a test and enables us to get shorter confidence intervals for a given confidence level.

Example 2.1

The problem. We return to the statistics books on my shelves. Numbers of pages for all 114 books are given in Section A7.1. In Example 1.2 we considered a random sample of 12, applied a sign test and did not reject the hypothesis that the median number of pages per book was 220. Would we reach the same conclusion if we doubled the sample size to 24? Below are numbers of pages, after ordering, for a random sample of 24 books from my collection:

153	166	181	192	244	248	258	264	296	305	305	312
330	340	356	361	395	427	433	467	544	551	625	783

Formulation and assumptions. If the population median, θ, is 220 then any book in a random sample is equally likely to have either more than or less than 220 pages (ignoring for a moment the possibility it has exactly 220 pages). The distribution of plus signs (number of books with more than 220 pages) is then binomial with $n = 24$ and $p = \frac{1}{2}$. We test the hypothesis

$$H_0 : \theta = 220$$

against

$$H_1 : \theta \neq 220$$

A two-tailed test is appropriate.

Procedure. There are four observations below 220, namely 153, 166, 181, 192. This implies 4 minus and 20 plus signs in our sample of 24. We need the binomial probabilities for $n = 24$ and $p = \frac{1}{2}$ given in Table 2.1 to see if this is significant. In view of symmetry of the binomial distribution where $p = \frac{1}{2}$, we need only record probabilities for $r \leqslant 12$: the probability of $24 - r$ plus signs equals the probability of r plus signs. This also means we may carry out a two-tail test by considering the total number of either plus or minus signs, since if the number of plus signs puts us in the upper critical region, the number of minus signs is at the corresponding symmetric value in the lower critical region.

Table 2.1 Binomial probabilities, P, for $n = 24$, $p = \frac{1}{2}$

r	0	1	2	3	4	5	6	7	8	9	10	11	12
p	0.000	0.000	0.000	0.000	0.001	0.002	0.008	0.021	0.044	0.078	0.117	0.148	0.162

Using principles outlined in Section 1.2 we choose a critical region of nominal size 0.05 by selecting from each tail outcomes with total associated probability not more than 0.025 in that tail. Clearly this includes the values 0 to 6 inclusive (associated total probability 0.011). Because of symmetry the values 18 to 24 from the upper tail have the same associated probability.

We observe four minuses, so our outcome is in the critical region for a test at the nominal 5% level (actual level for the two-tail test is $(2 \times 0.011) \times 100\% = 2.2\%$).

Conclusion. We reject the hypothesis that the median number of pages is 220. The preponderance of plus signs suggests a higher median.

Comments. 1. The values 0 to 5 form the lower tail of a critical region for a two-tail test of nominal size 0.01 (actual size $2 \times 0.003 = 0.006$). It is usual, and desirable, to assert significance at the lowest permissible of three commonly used nominal values, 5%, 1% or 0.1%. These levels are sometimes said to imply a result is **significant, highly significant** or **very highly significant** respectively. Our result is significant at the 1% level, i.e. highly significant. For some nonparametric tests critical regions for significance at the 0.1% level are not readily available from tables.

2. We used a two-tail test. In Section 1.2.1 we pointed out circumstances where a one-tail test might be appropriate. A decision on whether to use a one- or two-tail test depends on the logic of the experimental situation and should be made before the data are collected – certainly before they are inspected. Only if there is good reason to believe that a one-tail test reflects the only viable alternative should it be used. What should one do if a one-tail test is used but the result strongly indicates 'significance' in the other direction to that postulated in the alternative? If this happens in a situation where the one-tail test is logically justified, it may reflect a 'random' low-probability outcome for the particular experiment. As indicated in Section 1.2.1, an upper one-tail test is best formulated as: test $H_0 : \theta \leqslant \theta_0$ against $H_1 : \theta > \theta_0$, but this must not be taken as an invitation to use one-tail tests as significance boosters when there is no prior reason to regard the likelihood of θ being less than θ_0 as virtually zero.

3. If we reject the null hypothesis, interest switches to what the true value of θ is likely to be. This is usually our ultimate and really useful goal. We consider this in Example 2.2.

4. We had full data on numbers of pages in this example, but we could have carried out the sign test if we had known only that 4 of the 24 books had less than 220 pages.

In the sign test, and most others, we base our conclusion on some function of the sample values; for the sign test it is the number of positive (plus) or negative (minus) deviations from the median specified in the null hypothesis. Any function of the sample values used in a test is called a **statistic**.

Especially in a large sample, one or more sample values may coincide with our hypothesized θ_0; i.e. 220 in Example 2.1. We cannot logically assign either a plus or a minus sign to this observation. Our recommended procedure is to ignore this zero difference and to consider our sample size to be reduced by one. In doing so we are rejecting a piece of evidence that strongly supports the null hypothesis but is useless in telling us anything about the direction of a possible alternative. Another possibility is to toss a coin and replace a zero by a plus if it falls heads and a minus if it falls tails; this 'random' procedure usually makes only a small difference but it has little to commend it. Some writers recommend assigning a plus or a minus to a zero in a way that makes rejection of H_0 less likely. This is an ultraconservative approach.

We need a composite set of tables to find the critical region for various n. We could tabulate the appropriate probabilities for individual numbers of successes (i.e. numbers of specified sign) as in Tables 1.1 and 2.1. A more convenient way is to give not the probability for each r, but the cumulative probability that we observe r or fewer successes. Table 2.2 does this for values of $r = 0$ to 12 in the special case $n = 24$ and is formed from Table 2.1 by replacing each entry in the latter by the sum of all entries at and to the left of the specified r. These totals, giving the probabilities of r or fewer successes, are called **cumulative binomial probabilities.**

Table 2.2 Cumulative binomial probabilities, $n = 24$, $p = \frac{1}{2}$

r	0	1	2	3	4	5	6	7	8	9	10	11	12
P	0.000	0.000	0.000	0.000	0.001	0.003	0.011	0.032	0.076	0.154	0.271	0.419	0.581

From this table we see that for a two-tail test at the nominal 5% level (where we require at most 0.025 probability in the lower tail), r must not exceed 6 for the lesser of the number of plus signs or minus signs. For a one-tail test of $H_0 : \theta \geqslant \theta_0$ against an alternative $H_1 : \theta < \theta_0$, when testing at the nominal 5% level the critical region is 0 to 7 (actual size 0.032) and for a one-tail test at nominal 1% level the critical region is 0 to 5. These apply to the number of plus signs. For a one-tail test of $H_0 : \theta \leqslant \theta_0$ against an alternative $H_1 : \theta > \theta_0$ the same critical regions apply for the number of minus signs.

Table A1 gives cumulative probabilities for r or fewer successes for the binomial distribution with $p = \frac{1}{2}$, for $6 \leqslant n \leqslant 20$. These enable us to determine easily critical regions for various nominal and actual significance levels for one- or two-tail tests. Even simpler tables are given by Neave (1981, p. 29) together with an explanation of their use. They cover values of $n \leqslant 100$, but give only nominal (not actual) significance levels. Readily available tables for many tests described in this book give only nominal significance levels; this is one (regrettable) reason why it is common to quote nominal rather than actual significance levels. For $n \leqslant 20$ one may also use Neave's cumulative binomial probability tables if actual levels are required.

Table A1, and indeed most published tables, are restricted in values of n. In Section 2.1.3 we examine an approximation that works well when $n > 20$.

2.1.2 A confidence interval

Example 2.2

The problem. Obtain a nominal 95% confidence interval for median book length based on the sign test using the sample data in Example 2.1.

Formulation and assumptions. We seek all values for θ_0, the true median value, that would not be rejected in a two-tail significance test at the 5% level.

Procedure. We use the argument developed in Section 1.3. In Example 2.1 we found the critical region for a 5% significance test with a sample of 24 to be 0 to 6 and 18 to 24, implying we reject with 6 or fewer plus and 18 or more plus signs. The data in Example 2.1 are ordered, and examining these it is clear that we would not reject any hypothetical median which is greater than 258 (as this would still give at least 7 minus) but less than 427 (as this would give at least 7 plus). Values less than 258 or greater than 427 on the other hand would give 6 or less minus and 6 or less plus respectively.

Conclusion. A nominal 95% (actual $100(1 - 0.022)\% = 97.8\%$) confidence interval is 258 to 427.

Comments. 1. The 95% confidence interval assuming normality and based on a t-distribution (see Section A4.1) is (291, 420); the shorter length reflects greater efficiency with a stronger assumption. We discuss tests for validity of a normality assumption in Section 3.1.3. Note that nominal and actual confidence levels are identical for the interval based on the t-distribution, whereas the actual level is usually higher than the nominal for that based on the sign test; so the intervals are not exactly comparable.

2. In Section 1.3.1 we showed that with a sample of 12 the sign test procedure gave a 95% confidence interval (156, 454), and the corresponding interval based on a normality assumption and the t-distribution was (227, 405). Doubling our sample size and using a sign test basis gives, as asserted in Section 1.2.3, as good a result (in terms of interval length) as an interval based on the smaller sample and a t-distribution. The intervals do not coincide as they are based on different samples as well as on a different theoretical approach.

2.1.3 A large-sample approximation

If $n > 20$ an approximation based on the normal distribution is usually adequate. For any binomial distribution, if n is reasonably large and p not too small (so that $np > 10$ is a useful but approximate guideline) and X is the number of successes then

$$Z = \frac{X - np}{\sqrt{(npq)}} \qquad (2.1)$$

where $q = 1 - p$, has an approximately normal distribution with mean 0 and standard deviation 1 (the standard normal distribution). Tables giving probabilities that Z takes values less than or equal to any specified value (i.e. of the cumulative distribution function) are widely available. For the sign test $p = q = \frac{1}{2}$ and (2.1) becomes

$$Z = \frac{X - \frac{1}{2}n}{\frac{1}{2}\sqrt{n}} \qquad (2.2)$$

The approximation is improved by a continuity correction which allows for our approximating a discrete variable taking the values $0, 1, 2, \ldots, n$ by a continuous one which may take any real value. If r is the greater of the number of plus signs or minus signs we subtract $\frac{1}{2}$ from r, i.e. put $X = r - \frac{1}{2}$ in (2.2). If r is the lesser of the number of plus signs or minus signs we add $\frac{1}{2}$ to r, i.e. put $X = r + \frac{1}{2}$.

Testing at the 5% significance level, we reject H_0 in a two-tail test if Z is greater than 1.96 or less than -1.96, i.e. if Z takes values outside the interval $(-1.96, 1.96)$. It is convenient in this case to refer to $|Z| \geqslant 1.96$ as the rejection criterion, where the vertical bars indicate we take the magnitude (positive value) of Z; also referred to as the modulus of Z. For a one-tail test at the 5% level we replace 1.96 by 1.64 (or -1.64, if the lower tail is relevant). Corresponding critical values for rejection at the 1% level are 2.58 (two-tail) and 2.32 or -2.32 (one-tail).

Example 2.3

The problem. Given the data in Example 2.1, test the hypothesis $H_0: \theta = 220$ using the large-sample normal approximation to the two-tail sign test.

Formulation and assumptions. As in Example 2.1, but we now use (2.2) with a continuity correction.

Procedure. We observe 4 minus and 20 plus signs, so $r = 20$; with a continuity correction, $X = 20 - \frac{1}{2} = 19.5$. Since $n = 24$, (2.2) gives $Z = (19.5 - 12)/(0.5 \times \sqrt{24}) = 3.06$, indicating significance at the 1% level. Indeed, tables of the normal distribution show that $\Pr(Z > 3.06) = 0.0011$. Because of symmetry this equals $\Pr(Z < -3.06)$. From Table 2.2 we see that $\Pr(r < 4) = \Pr(r > 20) = 0.001$. A more precise value for this probability is 0.0009, in reasonable agreement with the normal approximation, 0.0011.

Conclusion. We reject H_0 at the nominal 1% significance level.

Comment. We arrive at the same conclusion if we work with the lower tail, putting $X = 4 + \frac{1}{2}$.

2.1.4 Large-sample critical values and confidence intervals

We may rearrange equation (2.2) to get approximations to the critical value, r_0, say, of the lesser of the number of plus or of minus values. If we regard $X = r_0$ as unknown but fix $Z = -1.96$, then for a given n we may calculate r_0. We reject H_0 if r, the lesser of the number of pluses or minuses, is less than or equal to r_0. It is better to use a continuity correction, i.e. put $X = r_0 + \frac{1}{2}$ for the lesser number. Simple algebraic manipulation of (2.2) with $Z = -1.96$ then gives

$$r_0 = \tfrac{1}{2}n - 0.98\sqrt{n} - \tfrac{1}{2} \tag{2.3}$$

Substituting $n = 24$ in (2.3) gives $r_0 = 6.69$. Since r can only take integral values we reject H_0 if $r \leqslant 6$, in agreement with the exact criterion. Similar calculations can be used with other significance levels for one- or two-tail tests.

Once the critical value r_0 is determined, confidence intervals may be obtained as for the exact test.

2.1.5 A modified sign test

We developed the sign test for a population median. If the population is symmetric then the mean and median coincide; then it is also a test about the population mean. If the population is not symmetric we cannot use the sign test as a test about the mean, for if μ, say, is the mean of a non-symmetrically distributed random variable X then generally $\Pr(X < \mu) \neq \Pr(X > \mu)$ so the numbers of plus or of minus signs of deviations from the mean no longer have a binomial distribution with $p = \frac{1}{2}$.

We can use a modification of the sign test to test hypotheses about distribution quantiles other than the median. The kth quantile of a continuous distribution is defined as the value q_k of the random variable X such that $\Pr(X < q_k) \leqslant k$ and $\Pr(X > q_k) \leqslant 1 - k$. Clearly $q_{1/2}$ is the median. For continuous distributions q_k is unique; for discrete distributions special conventions are needed to define unique quantiles (see Section A1.2).

In particular, If $k = 0.1r$, where r is any integer between 0 and 9, q_k is the rth decile (tenth) and if $k = \frac{1}{4}r$ where $r = 1, 2$ or 3, q_k is the rth quartile (quarter). The second quartile is the median; so is the fifth decile.

We illustrate a sign test for quantiles for the case of a first quartile. The extension to other quantiles follows similar lines.

Example 2.4

The problem. A central examining body awards a percentage mark in each subject and publishes the information that 'three-quarters of the candidates achieved a mark of 40 or more'. One school entered 32 candidates of whom 13 scored less than 40. The president of the Parents' Association argues this is evidence that the school's performance is below national standards. The headmaster counters by claiming that if one took a random sample of 32 candidates it is not unlikely that 13 score less than the lower quartile mark, even though 8 in 32 is the national proportion. Is his assertion justified?

Formulation and assumptions. Formally, the headmaster's assertion may be tested using a hypothesis test of $H_0: q_{1/4} = 40$ against the alternative $H_1: q_{1/4} \neq 40$.

Procedure. We associate a minus with a candidate's mark below 40; a plus with each mark over 40. We have 13 minuses (and 19 pluses) for our sample of 32. If the first quartile value is 40 then the probability is $\frac{1}{4}$ that each candidate in a random sample has a mark below 40 (and thus is scored as a minus) and the probability is $\frac{3}{4}$ for a mark above 40 (scored as a plus). The distribution of minuses is therefore binomial with $n = 32$, $p = \frac{1}{4}$. Tables of exact probabilities for this distribution exist (but are less common than those for $p = \frac{1}{2}$). Neave (1981) gives such a table for values of $n \leqslant 20$. It is tedious, but not difficult, to work out exact probabilities using a recursive formula (see Section A1.1) with a pocket calculator – easy to do so with a computer. For a two-tail test when $n = 32$ the critical region of nominal size 0.05 consists of 3 or less minuses or 14 or more minuses. The actual size of this region is 0.041.

Conclusion. Having observed 13 minus signs we do not reject the null hypothesis that the population median is 40. In that sense the headmaster's assertion is justified.

Comments. 1. Recall that non-rejection of a hypothesis does not prove it true. It is only a statement that the evidence to date is insufficient to reject it. We may still be making an error of the second kind by non-rejection.

2. We used a two-tail test. Would a one-tail test be justified? Most statisticians would say no, unless there was further information to indicate that it was highly unlikely the school performance could be better than the national norm. For example, if it were known that most schools devoted three periods of tuition per week to the particular subject, but the school in question only devoted two, it could well be argued by the Parents' Association that lack of tuition was depressing pupil performance below the national level. A one-tail test at the 5% level rejects H_0 for 12 or more minuses.

3. Although we do not recommend the normal approximation for values of $np < 10$, it can be used (especially with a continuity correction) in the form (2.1) if one is prepared to sacrifice a little accuracy and is not too dogmatic about requiring exact levels of significance. In the above example $n = 32$, $p = \frac{1}{4}$,

whence $np = 8$ and $npq = 32 \times \frac{1}{4} \times \frac{3}{4} = 6$. Since the number of minus signs is above the expected number $np = 8$, we subtract the continuity correction of $\frac{1}{2}$. Thus

$$Z = (12.5 - 8)/(\sqrt{6}) = 1.84$$

Since $Z < 1.96$ we do not reject H_0 for a two-tail test. As with the more exact test we would reject H_0 with a one-tail test, for the 'critical' value of Z is then (p. 25) 1.64.

4. Note how the headmaster's claim was worded in terms of 'if one took a random sample'. Pupils from just one school are in no sense a random sample from all candidates. What our test is establishing is that the results for candidates from this school are not out of line with national results in the sense that they are the sort of result one could easily get with a random sample of that number of candidates from all entrants.

Test for the third quartile are symmetric with those for the first quartile if we interchange plus and minus signs.

2.1.6 A sign test for trend

Cox and Stuart (1955) suggested an ingenious use of the sign test to determine evidence of a monotonic trend, i.e. an increasing or decreasing trend. A straight line is the simplest form of monotonic trend and with simple assumptions about departures from a linear trend, least squares regression is often an appropriate method of making inferences. But a monotonic trend need not be linear; it may simply express a tendency for observations to increase subject to local or random irregularities. Consider a set of **independent** observations $x_1, x_2, \ldots, x_n$ ordered, say, in time. If we have an even number of observations $n = 2m$, say, we take the differences $x_{m+1} - x_1, x_{m+2} - x_2, \ldots, x_{2m} - x_m$. For an odd number of observations, $2m + 1$, we may proceed as above, omitting the middle value x_{m+1} and calculating $x_{m+2} - x_1$, etc. If there is an increasing trend we would expect most of those differences to be positive, whereas if there were no trend and the observations differed only by random fluctuations about some median these differences (in view of the independence assumption) are equally likely to be positive or negative. A preponderance of negative differences suggests a decreasing trend.

This implies that under the null hypothesis of no trend, the plus (or the minus) signs are distributed with a binomial distribution with parameters m and $p = \frac{1}{2}$.

Example 2.5

The problem. The US Department of Commerce publishes estimates obtained from independent samples of the mean mileages covered by various classes of vehicle in the United States each year. The figures for cars and trucks (in thousands of miles) are given below in order for each of the years 1970–83. In either case is there evidence of a monotonic trend?

Cars	9.8	9.9	10.0	9.8	9.2	9.4	9.5	9.6	9.8	9.3	8.9	8.7	9.2	9.3
Trucks	11.5	11.5	12.2	11.5	10.9	10.6	11.1	11.1	11.0	10.8	11.4	12.3	11.2	11.2

Formulation and assumptions. As the figures for each year are based on independent samples we may use the Cox–Stuart test for trend. Without further assumptions a two-tail test is appropriate as any trend may be in either direction.

Procedure. For cars the relevant differences are $9.6 - 9.8$, $9.8 - 9.9$, $9.3 - 10.0$, $8.9 - 9.8$, $8.7 - 9.2$, $9.2 - 9.4$, $9.3 - 9.5$, and all are negative. From Table A1 for $n = 7$ we find that the 0.05 critical region (actual size 0.016) consists of 0 or 7 plus signs.

Conclusion. There is evidence of a monotonic trend. Clearly this is downward.

Comments. 1. For trucks the corresponding differences have the signs $-$, $-$, $-$, $-$, $+$, $+$, $+$. A moment's reflection shows 3 plus signs and 4 minus signs (or 3 minus and 4 plus) provide the strongest possible evidence to support the null hypothesis of no monotone trend.

2. Periodic trends are common. For example, in the northern hemisphere mean monthly temperature tends to increase from January to July, then decrease from July to January. A Cox–Stuart test as used above might miss any such trend or even show it as a monotonic trend depending on the period of records. Do you think there might be some non-monotonic trend for the US truck mileage data? Conover (1980, Example 5, p. 137) gives an interesting example of adapting the Cox–Stuart test to detect periodic trend in certain circumstances by reordering the data.

3. If the same samples of cars and of trucks had been used each year, the independence assumption would not hold and any inferences would only be valid, if at all, for the vehicles in that sample, for anything atypical about the sample would influence all observations. With independent samples in each year anything atypical about the sample in one year will be incorporated in the random deviation from trend for that year. If the samples are not random there is the possibility of bias and the samples may not properly reflect population characteristics of all cars in the US.

2.1.7 Extending the sign test beyond one sample

In Chapter 4 we consider situations where we have two related samples. The sign test plays an important role in this situation; the median test extends immediately, and the same method is also used in a class of tests called McNemar tests.

2.2 INFERENCES ABOUT MEDIANS BASED ON RANKS

The sign test for a median uses little of the information in a data set like that of numbers of pages in Example 2.1. We note only whether each observation is above or below the median specified by H_0. At the same time it is applicable to random samples that are only ordinal values from virtually any population and allows valid inferences about that population median. It is even valid if we only have counts of numbers of items above or below the θ specified in H_0. To obtain confidence intervals we do, however, need measurement values.

If we make the further assumption that the population distribution is symmetric then the point of symmetry is the population median or mean (they coincide) and we can make inferences in a way that makes greater use of measurement data.

We assume our population distribution is continuous. This means there is (in theory) a zero probability of two sample values coinciding. This is an unrealistic assumption for much real data, but it is a complication we can cope with by modification of the test we now describe.

2.2.1 The Wilcoxon signed rank test

Given a sample of n independent measurements, instead of just noting the sign of departures from the median or mean specified in H_0 we may find also the magnitude of each departure. If H_0 specifies the true median or mean of a **symmetric** distribution, departures of any given magnitude are equally likely to be positive or negative; e.g. if θ is the median of a symmetric distribution a value between, say, 4 and 5 units above θ has the same probability as a value between 4 and 5 units below θ. Thus, if we arrange all sample deviations from the hypothesized median, θ_0, in order of **magnitude** and replace them by their ranks (1 for the smallest deviation and n for the largest) and then place a negative sign before ranks corresponding to values below the median, we expect a good scatter of positive and negative ranks if θ_0 is the true median. If we took the sum of all positive ranks and the sum of all negative ranks we would not expect those sums to differ greatly.

A high or low value of the sum of the positive (or negative) ranks relative to the sum of the negative (or positive) ranks implies θ_0 is unlikely to be the median or mean. These ideas are incorporated in the **signed rank test** proposed by Wilcoxon (1945).

Example 2.6

The problem. In our preliminary discussions in Chapter 1 we considered numbers of pages of a random sample of 12 from 114 statistics textbooks. They were

$$126 \quad 142 \quad 156 \quad 228 \quad 245 \quad 246 \quad 370 \quad 419 \quad 433 \quad 454 \quad 478 \quad 503$$

In that chapter we accepted the hypothesis $H_0: \theta = 220$ using a two-tail sign test. Making now an additional assumption that the numbers of pages are symmetrically distributed about the true median, use the Wilcoxon signed rank test, for a two-tail test of the same hypothesis.

Formulation and assumptions. We arrange deviations from 220 in order of magnitude and associate with each rank the sign of the corresponding deviation. We then compare the lesser of the sum of positive and negative ranks with tabulated values to assess significance.

Procedure. Subtracting 220 from each sample entry we get the differences $-94, -78, -64, 8, 25, 26, 150, 199, 213, 234, 258, 283$. Rearranging these in order of increasing magnitude and retaining the signs we get $8, 25, 26, -64, -78, -94, 150, 199, 213, 234, 258, 383$; the corresponding 'signed' ranks are $1, 2, 3, -4, -5, -6, 7, 8, 9, 10, 11, 12$. Clearly the sum of the negative ranks is $S_n = 15$ and this is less than the sum of the positive ranks, S_p. We use Table A2 or the tables in Neave (1981, p. 29) to see if this is significant. Entries in these tables have been calculated by a generalization of a method we outline in Section 2.2.2 below. Table A2 shows that if $n = 12$ we reject H_0 in a two-tail test at the 5% significance level only if the lesser of S_n, S_p is less than or equal to 13. We found $S_n = 15$.

Conclusion. We do not reject H_0 at the nominal 5% significance level.

Comments. 1. Is it reasonable to assume that numbers of pages in a book are symmetrically distributed about the mean? Most books are between about 100 and 500 pages in length and if the true mean were about 300 it may not be unreasonable to expect an approximately symmetric distribution. However, there are likely to be a few books of 600, 700 or more pages in a large collection of statistics texts, so there is at least some asymmetry. If the number of such books is relatively small the Wilcoxon test should not be seriously misleading.

2. We assumed continuity in our preliminary discussion. Clearly numbers of pages are integers, but in view of the wide spread this is something like a rounding effect that does not seriously undermine the validity of the test.

2.2.2 Theory of the Wilcoxon signed rank test

We illustrate this for a small sample of seven observations. Intuitively we expect the sums of negative and positive signed ranks of deviations to be approximately equal if H_0 is true, whereas there will be an appreciably higher value of one sum compared with the other if H_0 is untrue. In an extreme case (if all sample values are above the value specified in H_0) all ranks will be positive, so $S_n = 0$; then for seven observations $S_p = 1 + 2 + \cdots + 7 = 28$, $S_n = 0$. For any mixture of positive and negative ranks when $n = 7$, $S_n = 28 - S_p$.

More generally, for n observations with no ties and no observation equal to θ_0, $S_n = \frac{1}{2}n(n+1) - S_p$, since the sum of the ranks $1, 2, 3, \ldots, n$ is $\frac{1}{2}n(n+1)$. This relationship implies that the probability that the positive ranks sum to S_p is the same as the probability that the negative ranks sum to $\frac{1}{2}n(n+1) - S_p$.

Each rank may have either a positive or a negative sign attached to it, so signs may be attached to 7 ranks in $2^7 = 128$ different ways. If signs are allocated entirely at random (e.g. by toss of a coin) the probability that all will be positive (i.e. $S_n = 0$) is $1/128$. If only rank 1 is negative, the sum of negative ranks is 1, again with probability $1/128$. If only rank 2 is negative, the sum of negative ranks is 2, again with probability $1/128$. If ranks 1 and 2 are both negative, or if only rank 3 is negative, in either case $S_n = 3$; since $S_n = 3$ can occur in two ways it has probability $2/128$. In view of the symmetry, the probabilities that $S_n = 26$, 27 or 28 are each $1/128$ and the probability that $S_n = 25$ is $2/128$. These extreme values are relevant to a critical region for a two-tail test; they are improbable if the population mean actually has the value hypothesized under H_0. A nominal 0.05 critical region for the two-tail test (actual size 0.047) is found by taking the rank sum values 0, 1, 2 from the lower tail and 26, 27, 28 from the upper tail. In view of the symmetry this implies rejection if the lesser of S_p, S_n does not exceed 2.

For a one-tail test at the nominal 5% level we reject if the sum of the positive or negative ranks (as appropriate) does not exceed 3; the actual size is then $5/128 = 0.039$, implying an actual 3.9% significance level.

2.2.3 A large-sample approximation

For sample size $n > 20$ a normal approximation to the Wilcoxon test works well. Denote the **magnitude** of the lower of the positive or negative rank sums by S; it can be shown that S has mean $(n+1)/4$ and variance $n(n+1) \cdot (2n+1)/24$, whence

$$Z = \frac{S + \frac{1}{2} - \frac{1}{4}n(n+1)}{\sqrt{[n(n+1)(2n+1)/24]}} \tag{2.4}$$

has a standard normal distribution; the $\frac{1}{2}$ in the numerator is a continuity correction. Critical values of Z for a two-tail test are $Z < -1.96$ at the 5% level and $Z < -2.58$ at the 1% level; the corresponding one-tail levels are -1.64 and -2.33.

In Example 2.6 we only had a sample of 12, but substituting $S = 15$ in (2.4) gives $Z = -1.84$, so we would not reject the null hypothesis; consistent with the finding in our exact test. Just as we did for the large-sample normal approximation for the sign test, we may use (2.4) to find a critical value of S for significance for a given n by computing the S corresponding to the appropriate Z. Bearing in mind that the approximation may not be very good with $n = 12$ we nevertheless illustrate the procedure for that case at a 5% level in the two-tail test where $Z = -1.96$ is the critical value. Rearranging (2.4) with S unknown, $Z = -1.96$ and $n = 12$ we find $S = 13.51$. Since S can take only integral values we reject for a lower sum 13 or less; in agreement with tables for the exact test. For a one-tail test at the 5% level we set $Z = -1.64$, when, for $n = 12$, we find $S = 17.59$, indicating rejection for a lower sum less than or equal to 17, again in accord with Table A2. Thus at the 5% significance level the approximation is good for n as low as 12; we examine the approximation at the 1% level in Exercise 2.8.

2.2.4 Wilcoxon with ties

In theory, when we sample from a continuous distribution the probability of getting tied observations is zero: so is the probability of getting a sample value exactly equal to the population mean or median. In practice observations do not have strictly continuous distributions because of rounding or limited precision of measurement. We measure lengths to the nearest centimetre or millimetre, weights to the nearest kilogram, gram or milligram; the number of pages in a book in complete pages, although chapter layout often results in part-pages of text. These practical limitations may produce rank ties or zero differences. If they do, the exact distribution of S, the lower rank sum, becomes complicated and requires recomputation for different numbers of ties and ties in different position in the rank order.

One practical solution is to modify the normal approximation in Section 2.2.2 after adjusting our scoring method for ranks. If one or more observation coincides with the mean specified under H_0 (i.e. its deviation from that hypothesized mean is zero) we record these as zero rank. Denote by d_0 the number of such zeros. If two or more ranks are equal in magnitude (but not necessarily of the same sign) we replace them by what is called the mid-rank; e.g. if seven observed deviations are 3, 4.7, -5.2, 5.2, 7, 7, -7, we rank 3 as 1, 4.7 as 2, the 'tied' -5.2 and $+5.2$ are each allocated the mean of 3 and 4, i.e. 3.5, and the values 7, 7, -7 are given the mean of the remaining ranks 5, 6, 7, i.e. each is ranked 6, with appropriate sign. Our final signed rankings are thus 1, 2, -3.5, 3.5, 6, 6, -6. Note that the sum of the absolute ranks or rank magnitudes (i.e. ignoring minus signs) is still 28, as it would be for a straight ranking 1 to 7. S_n (the sum of the negative ranks) is now 9.5.

For zero deviations (observations coincident with θ_0) allocation of ranks is more complicated. Temporarily, zero is given formally the rank 1, as it represents the smallest deviation, but since zero is neither positive nor negative we replace that 1 by zero before summing the ranks. Thus differences of 0, 3, $-7, 11, -13, 16, -18$ are ranked without signs as 1, 2, 3, 4, 5, 6, 7 respectively. We may then allocate signs and get adjusted values by multiplying each unsigned rank by $+1$ if the corresponding difference is positive, by 0 if it is zero and by -1 if it is negative, giving in this case the signed ranks 0, 2, $-3, 4, -5$, 6, -7, with $S_n = 15$. The sum of the absolute ranks is now 27 rather than 28.

For each set of tied ranks accorded a mid-rank, we denote the number of observations in the ith such tie by d_i, $i = 1, 2, \ldots, r$ if there are r sets of different ties. If there are no ties in rank, we formally define $d_i = 1$.

Example 2.7

Suppose we have the following ordered deviations in a Wilcoxon signed rank test:

$$0, 0, 3, -7, 9, 9, -11, 11, 11, 14, 16, 17, 17, 18$$

We allocate the following unsigned ranks:

$$1.5, 1.5, 3, 4, 5.5, 5.5, 8, 8, 8, 10, 11, 12.5, 12.5, 14$$

To get the appropriate signed ranks we multiply by $+1, 0, -1$ for a positive, zero, or negative difference, whence the signed ranks become

$$0, 0, 3, -4, 5.5, 5.5, -8, 8, 8, 10, 11, 12.5, 12.5, 14$$

whence $S_n = 12$. Here $d_0 = 2$; and there are two ties of two ranks and one tie of three ranks for which we put $d_1 = d_2 = 2$, $d_3 = 3$.

Our large-sample test is a modification of (2.4). As it is now possible for S to take fractional values we omit the continuity correction of $\frac{1}{2}$ in the numerator and we replace the mean $\frac{1}{4}n(n+1)$ by $\frac{1}{4}[n(n+1) - d_0(d_0+1)]$ and modify the denominator by replacing the square root of $n(n+1)(2n+1)/24$ by the square root of

$$[n(n+1)(2n+1) - d_0(d_0+1)(2d_0+1)]/24 - \sum[d_i^3 - d_i]/48$$

Note that when $d_0 = 0$ (no zeros) and all $d_i = 1$ (no tied ranks) and the continuity correction is restored we regain (2.4).

Essentially the normal approximation is a large-sample result. If there are many ties in samples of less than 20, one should be wary about using the approximation and indeed one may be safer to use the standard signed rank test with mid-ranks but otherwise ignoring ties. However, significance levels will no longer coincide exactly with those for the no-ties case. Hopefully in small samples we will usually get only a few ties; if we use mid-ranks the effect on the exact test is then small.

For only a few ties or zeros the adjustments proposed above do not greatly alter the value of Z in (2.4) as the following example shows. If the numbers of ties or zeros are appreciable the effect may become quite marked. It does no harm to use the corrections in all cases at least until one has sufficient experience to judge the likely effect. A number of writers give essentially equivalent correction methods for ties formulated in slightly different ways; essentially these involve direct calculation based on ranks or mid-ranks of the standard deviation in the denominator of (2.4); see comment 2 on Example 2.8 for more details.

Example 2.8

The problem. Use the Wilcoxon signed rank test to test the hypothesis $H_0: \theta = 220$ against $H_1: \theta \neq 220$ for the data on numbers of pages in Example 2.1; i.e. page numbers for a sample of 24:

153 166 181 192 244 248 258 264 296 305 305 312
330 340 356 361 395 427 433 467 544 551 625 783

Formulation and assumptions. To justify the Wilcoxon test, we make an additional assumption of a symmetric distribution about the population mean or median. We determine the lesser of S_p or S_n after ranking deviations from 220 and use the modified version of (2.4).

Procedure. The actual signed deviations from 220 arranged in ascending order of magnitude are:

24 − 28 28 38 − 39 44 − 54 − 67 76 85 85 92
110 120 136 141 175 207 213 247 324 331 405 563

giving signed ranks (mid-ranks for ties) 1, − 2.5, 2.5, 4, − 5, 6, − 7, − 8, 9, 10.5, 10.5, 12, 13, 14, 15, 16, 17, 18, 19, 20, 21, 22, 23, 24. Clearly $S_n = 22.5$ is less than S_p. With two ties each involving two observations we have $d_1 = d_2 = 2$ and $n = 24$. Using the modifications given in this section, only the standard deviation is modified (i.e. the denominator of (2.4)) and the continuity correction is omitted in the numerator. It is easily verified that our test statistic using this modified form of (2.4) is

$$Z = \frac{22.5 - 150}{\sqrt{(1225 - 0.25)}} = - 3.54 \qquad (2.5)$$

Conclusion. We reject H_0 at the 0.1% significance level since normal distribution tables show the critical value for doing so in a two-tail test is $Z = - 3.29$.

Comments. 1. The correction in the denominator for a pair of ties is trivial in the above example and has little effect. Indeed, only if n is small (when the normal approximation is of dubious validity), or the number of ties is very large, is the correction appreciable. Many writers advocate using S_n or S_p and Table A2 or the standard approximation in (2.4) for only a few ties. This may not be seriously misleading, but actual, and sometimes even nominal significance levels, may no longer hold. This is discussed by Lehmann (1975, p. 130).

2. An alternative form for the adjustment for ties is to replace the denominator of (2.4), i.e. $\sqrt{[n(n+1)(2n+1)/24]}$ by $\sqrt{[\sum_i(r_i^2)/4]}$ where r_i, $i = 1, 2, \ldots, n$ is the ith allocated rank (or tied rank). It is a matter of personal preference whether to calculate the denominator by this method or by using the d_i. For a computer program it is perhaps easier to use the form $\sqrt{[\sum(r_i^2)/4]}$ which is valid whether or not there are ties since it reduces to $\sqrt{[n(n+1)(2n+1)/24]}$ when there are no ties.

3. In Example 2.1 applying the sign test to the same data we established significance only at a nominal 1% level. More information has sharpened our inference; we assert more confidently that the median is unlikely to be 220.

2.2.5 An alternative test statistic

We based our tests on the lesser of S_p, S_n. Some people use an alternative statistic $W = |S_p - S_n|$, i.e. the magnitude of the difference between the sums of positive and negative ranks. Since $S_p + S_n = \frac{1}{2}n(n+1)$, it is easily shown that $W = \frac{1}{2}n(n+1) - 2S_n$ in the case where $S_n < S_p$, so that there is a $1:1$ correspondence between values of S_n and W. Different tables (which can be derived from Table A2) are needed if W is the test statistic. It is more usual to use the lesser of S_p or S_n as we have done for an exact test. The normal approximation corresponding to W has the simple form

$$Z = W \Big/ \left[\sqrt{\sum_i (r_i^2)} \right] \tag{2.6}$$

since W has expectation zero under the null hypothesis. If there are no ties the denominator reduces to $\sqrt{\{n(n+1)(2n+1)/6\}}$.

2.2.6 Confidence intervals based on signed ranks

A symmetry assumption for the population distribution implies the mean and median coincide, so confidence intervals will be valid for both mean and median.

Calculation of confidence intervals for many nonparametric methods lacks the simplicity of hypothesis testing, but the procedure, once mastered, is simple in principle and easily programmed for a computer if a package program is not available.

The theory is given by Lehmann (1975, pp. 182–3). The basic idea is to take every pair of observations and calculate the mean of each pair as an estimate of the population mean. Since each sample value is equally likely to be above or below the population mean and equally likely to come from either of a pair of intervals symmetrically placed about that mean, these are sensible estimates. We expect some to be overestimates and some to be underestimates; it is intuitively reasonable to reject some of the smallest and some of the largest. Our intuition does not tell us how many to reject.

An interesting point is that if we include as well as different pairs (of which there are for n observations, $\frac{1}{2}n(n-1)$), the estimate obtained by taking the mean of each observation with itself (of which there are n), the total number of estimates is

$$\frac{1}{2}n(n-1) + n = \frac{1}{2}n(n+1)$$

This is also the sum of the ranks 1 to n. In Section 2.2.2 we showed that we reject H_0 when using a two-tail test if the lesser, S, of S_p and S_n is in the sub-set of smallest rank sums with associated total probability $\frac{1}{2}\alpha$ when testing at the $100\alpha\%$ significance level. Table A2 gives the appropriate critical values of S for which a result is just significant. The relevant confidence interval theory shows we get a $100(1-\alpha)\%$ confidence interval if we use the same value of S to indicate the number of smallest of the $\frac{1}{2}n(n+1)$ paired means we should omit. This gives a lower limit to the confidence interval. The upper limit is obtained by excluding the corresponding number of highest estimates. The procedure is illustrated in Examples 2.9 and 2.10. If we arrange the computations systematically to obtain limits we only need compute the highest and lowest paired values after checking first from Table A2 how many are to be excluded. An appropriate point estimate is the Hodges–Lehmann estimator, the median of all $\frac{1}{2}n(n+1)$ paired estimates, so if we require this we must compute further paired estimates (see comment 4 on Example 2.10 below). We frequently meet estimators of the Hodges–Lehmann type in nonparametric methods. They have important **robustness** properties (see Section 9.1)

Example 2.9

The problem. Warning devices fitted to aircraft give a visual or audible signal of impending danger when the speed of an aircraft is still some 5 knots above stalling speed. Seven of these devices were selected at random from a long production run and tested in a machine that simulates aircraft conditions. The speeds in knots above the nominal stalling speed at which the devices were triggered were

$$4.8, 5.3, 6.1, 4.2, 4.9, 5.9, 5.1$$

Obtain a 95% confidence interval for the mean excess over stalling speed at which the devices from the production run are triggered.

Formulation and assumptions. We see from Table A2 (and indeed established in section 2.2.2) that for samples of seven we reject H_0 at the nominal
5% level if $S \leqslant 2$. Thus we reject the two smallest and the two largest paired
means (remembering these may include observations paired with themselves,
i.e. the individual observation values).

Procedure. It is easy to find the two largest and the two smallest paired
values if we first order the observations, giving

$$4.2 \quad 4.8 \quad 4.9 \quad 5.1 \quad 5.3 \quad 5.9 \quad 6.1$$

Clearly the smallest observation corresponds to the smallest 'paired' mean.
The next smallest mean is given by the mean of 4.2 and 4.8, i.e. $\frac{1}{2}(4.2 + 4.8) =$
4.5; the next smallest is $\frac{1}{2}(4.2 + 4.9) = 4.55$. This is smaller than the second
observation 4.8, which is even larger than the mean $\frac{1}{2}(4.2 + 5.1) = 4.65$. Clearly
all remaining means are larger. The top three largest means are easily
evaluated as 6.1, $\frac{1}{2}(5.9 + 6.1) = 6.0$, and 5.9.

Conclusion. We omit the two smallest means of 4.2 and 4.5; the lower
confidence limit is given by the next smallest mean of 4.55. Similarly, striking
out the two highest means of 6.1 and 6.0 gives the upper confidence limit 5.9.
Our nominal 95% confidence interval is thus (4.55, 5.9).

Comment. The number of means to be calculated increases rapidly as n
increases. For example, when $n = 20$ we reject (see Table A2) the 52 lowest and
52 highest means to establish a 95% confidence interval. To do this without
error requires the more systematic approach outlined in Example 2.10.

Example 2.10

The problem. Using the sample of numbers of pages of 12 books in
Example 2.6, i.e.

$$126 \quad 142 \quad 156 \quad 228 \quad 245 \quad 246 \quad 370 \quad 419 \quad 433 \quad 454 \quad 478 \quad 503$$

calculate 95% and 99% confidence intervals for mean number of pages.

Formulation and assumptions. Table A2 indicates that the critical value of S
(here the numbers of smallest or largest means to be rejected) is 13 for a 95%
interval and 7 for a 99% interval.

Procedure. To facilitate calculation and accuracy (and to indicate the way
calculations might be programmed for a computer) we set these out in
Table 2.3. The entries are made in the following order. The first row and
column contain the ordered data given above.
Entries in the body of the table are on or to the right of the main diagonal

Table 2.3 Paired means for confidence limits

	126	142	156	228	245	246	370	419	433	454	478	503
126	126	134	141	177	185.5	186	248					
142		142	149	185	193.5	194						
156			156	192	200.5							
228				228								
245												
246												
370											424	436.5
419								426	436.5	448.5	461
433								433	443.5	455.5	468
454									454	466	478.5
478										478	490.5
503											503

and are the means of the observations at the head of the corresponding row or column. Thus main diagonal entries correspond to the observations themselves. We first write 126 at the top left of the main diagonal and then the next diagonal entry 142. We now proceed to calculate entries in the first row, i.e. $134 = \frac{1}{2}(126 + 142)$, $141 = \frac{1}{2}(126 + 156)$, $177 = \frac{1}{2}(126 + 228)$, etc. We note at this stage that 177 is bigger than any other entry to date. We next compute further means in the second row until we get one that equals or exceeds 177, i.e. we enter 149, 185. We now enter row 3 and calculate means until we get one larger than any yet calculated, i.e. 156, 192. We note that the first entry in the next row would be 228; bigger than any yet calculated. We now return to row 1 calculating any further entries that are smaller than those already obtained in other rows, continuing this process in following rows until we get the pattern in the top left of Table 2.3.

Note in particular that the entries increase as we move across rows from left to right and down any column. Thus the smallest entries are in this top left corner. It is easy to check that the 13 smallest entries (as required for a 95% confidence interval) are those less than or equal to 194. The next smallest entry is 200.5. This gives the lower confidence limit. The bottom right entries in the table are formed in similar manner, starting with the last row; these are the largest entries. Entries decrease as we move up a column or across a row from right to left. Eliminating the 13 largest means gives an upper confidence limit of 433.

Conclusion. A 95% confidence interval for the mean is (200.5, 433). Since a 99% interval is obtained by eliminating the 7 largest and 7 smallest means it is easily seen from Table 2.3 that the 99% confidence interval is (185, 455.5).

Comments. 1. Note that not only do the paired means increase as we move from left to right across rows or down columns in Table 2.3 but that there is a constant difference between entries in each column for pairs of rows. For example, entries in the third row that have an entry above them in the second row are all 7 greater than that second-row entry. This facilitates computation of entries.

2. In Section 1.3 we obtained a 95% confidence interval of (156, 454) based on the sign test, somewhat larger even than the nominal 99% confidence interval above. Our 95% interval of (200.5, 433) is still larger than the 95% interval based on the *t*-distribution with an assumption of normality given in Section 1.3; i.e. (227.4, 405.9).

3. The procedure used in this example forms a good basis for a computer program for confidence intervals. The computer is well suited to systematic and dreary calculations like finding means of pairs of observations.

4. To obtain the Hodges–Lehmann point estimate of the mean (or median) we need the median of all entries in a completed tableau like that of Table 2.3. Using the pattern of increases in rows and columns this can be found without completing the whole table by working in a region near the median diagonal value, i.e. the median of all sample observations (see Exercise 2.3), but with a computer program it is straightforward to compute all entries in the table and obtain their median.

2.2.7 A graphical method for a confidence interval

Ingenious graphical methods have been evolved for determining confidence intervals of the Wilcoxon type. For interest, a brief description of one is given here in the form of an example, but the method is laborious compared with the use of a suitable computer program based on the method described in the last section.

Example 2.11

The problem. Use a graphical method to obtain a 95% confidence interval based on the data in Example 2.9, i.e. given the seven speed differences

$$4.2 \quad 4.8 \quad 4.9 \quad 5.1 \quad 5.3 \quad 5.9 \quad 6.1$$

Formulation and assumptions. The graphical method is described in the procedure and gives the same solution as that obtained numerically in Example 2.9.

Procedure. We describe the method with the aid of Figure 2.1. It is best to use graph paper in practice.

We draw a vertical line AB and on it mark, on an appropriate scale, points

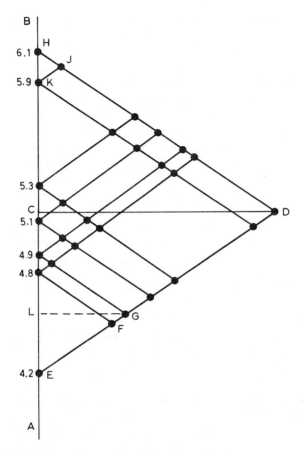

Figure 2.1 Graphical method for a confidence interval.

corresponding to each observation. The point C lies half-way between 4.2 and 6.1, the smallest and largest observations, represented by E and H on the graph. CD is perpendicular to AB and of any convenient length. D is joined to E and H. Through each point corresponding to the five remaining observations we draw lines parallel to HD and ED. At all intersections of these sets of parallel lines, including the data points, we place a clear dot. If we were to drop perpendiculars to AB from each dot (but there is no need to do this) the points where these met AB would give the paired means.

We saw in Example 2.9 that we drop the two largest and two smallest paired means. These correspond to the points labelled H, J, F, E in Figure 2.1. The confidence limits are then given by vertical scale values corresponding to the next data points working inwards form top and bottom. These are G (with scale value marked L) and K, and give the values 4.55 and 5.9 as the confidence limits.

Conclusion. A 95% confidence interval for the population median or mean is (4.55, 5.9).

Comments. Since there are $\frac{1}{2}n(n + 1)$ paired means (including the observed values), i.e. 28 when $n = 7$, the median of these pairs is mid-way between the 14th and 15th largest. Counting from the bottom in Figure 2.1 it is easily verified that the median of all pairs has the value 5.125. This is the Hodges–Lehmann point estimator of the population median or mean. In this case it is not very different from the median of all seven observations, i.e. 5.1, or the mean of those observations, 5.19, indicative of near symmetry.

2.3 OTHER LOCATION ESTIMATORS

There are other location tests and estimators closely related to the Wilcoxon signed rank estimator and that have intuitive appeal. However, in practice they are more difficult to calculate and usually give little, if any, gain in efficiency. We describe some of these in Section 3.3.

2.4 FIELDS OF APPLICATION

Here are a few illustrative examples from various fields of the type of problem to which methods in this chapter may be relevant.

Insurance

The median of all motor claims paid by an insurance company in 1987 is calculated. It might be £422. Early in 1988 the company suspects successful claims are higher. To test this, or estimate how much higher they might be, they take a random sample of 25 claims and test whether there is evidence of an increase. Since the distribution of claims may be rather skew a sign test would be appropriate. For a hypothesis test they need only note if each claim is above or below £422. Actual values would be needed for an interval estimate, i.e. a confidence interval.

Medicine

The median pulse rate for boys of a certain age prior to physical training is known. If pulse rate is taken for a sample after exercise the sign test could be used to test for a shift in median. Would you consider a one- or a two-tail test appropriate? If the original data for a large group of children indicated a symmetrical distribution of pulse rates the Wilcoxon test may be appropriate, but some caution is needed as symmetry prior to exercise need not imply symmetry afterwards; e.g. rate increases might be relatively higher for people with rest pulse rates above the median than is the case for those with rest pulse rates below the median prior to exercise, and this could give rise to skewness.

The researcher asking questions about change will probably know if this is likely to be so.

Engineering

The median noise levels under the flight path into an airport might be known for planes with a certain engine type (the actual level will vary from plane to plane and day to day depending on meteorological conditions, the precise height it passes over the measuring point, etc.). If the engine design is modified and a sample of measurements made under similar conditions to that for the old engine, an indication can be gained as to whether there is a noise reduction. A one-tail test would be appropriate if it were clear the modification could not increase noise, but might reduce it. We are unlikely to be able to use a true random sample here, but the first 40 approaches by aircraft using the new engines (if taken in a wide range of atmospheric conditions) might broadly reflect the characteristics of a random sample.

Biology

Heartbeat rates for female monkeys of a particular species in one locality may have a symmetric distribution with known median. Given heartbeat rates for a sample of females of the same species from another locality the Wilcoxon test could be used to detect a shift or to obtain confidence limits for the true mean value. With the further assumption of normality, normal theory procedures might be appropriate.

Physics

Specimens of metal are heated and their hardness measured on independent samples at $5°$ intervals over a $90°$ temperature range. The Cox–Stuart test might be a useful preliminary indicator of a monotone relationship between hardness and temperature.

Education

A standard test of numerical skills for 12-year-old boys is widely used. The median mark is established as 83. A new method of teaching such skills is introduced at one school and marks recorded for a class of 42. The large-sample approximation to the Wilcoxon test could be used to detect a change in median associated with the new method if symmetry could be assumed. If not, a sign test is preferable. There may be some reservation about regarding the test group as a random sample from a specific population; see comment 4 on Example 2.4.

Management

Records are used to obtain the mean number of days absent from work for all employees in a large factory in a given year. The number of days absent for a random sample of 25 is noted for the following year. To test for change the Wilcoxon test is appropriate if symmetry is assumed. If symmetry cannot be assumed the sign test is available. Is symmetry likely to be a reasonable assumption for data on absences?

Geography and environment

Frequency and duration of natural phenomena are often recorded. At a proposed airport site median cloud cover might be estimated by taking over a long period regular observations of cloud cover at fixed times each day. The site might be rated unsuitable if indications were that the median cover was high. A confidence interval for the median estimate would be useful.

EXERCISES

2.1   The first application listed in Section 2.4 involved insurance claims. Suppose, as in that example, that in 1987 the median claim was £422. A random sample of 15 claims from a large batch received in 1988 showed amounts (in £) of

386   394   421   422   457   581   633   790   1230   1250   1560   1770
1903   4399   5600

What test do you consider appropriate for a shift in median?
Would a one-tail test be appropriate?
Obtain a 95% confidence interval for the median based upon these data.
   In Section 2.1.3 we recommended the normal approximation for the sign test only if $np > 10$. Try it in this example and see if you reach a similar conclusion to the exact sign test.

2.2   For the problem in Example 2.8 show that using formula (2.6) leads to the same conclusion as we arrived at in that example.

2.3   For the data in Example 2.10 obtain the Hodges–Lehmann point estimator of the median.

2.4   Obtain the 99% confidence interval for the data in Example 2.10 using the graphical method described in Section 2.2.7.

2.5   The weight losses in kilograms for 16 overweight ladies who have been on a diet for two months are as follows:

4   6   3   1   2   5   4   0   3   6   3   1   7   2   5   6

The firm sponsoring the diet says in its advertisements 'Lose 5 kg in two months'. Tackled in a consumer watchdog radio programme on what this means, a spokesman says this is what a fat lady will lose 'on average'. You may be unclear about what he means by 'on average' but assuming it is reasonable to regard the sample of 16 as effectively a random sample, do you think it indicates a median weight loss of 5 kg? Test this without any assumption of population symmetry.

What would be a better test with an additional assumption of symmetry? Carry out the latter test.

Assuming symmetry, find nominal 95% confidence limits for the mean weight loss. Obtain also the confidence limits that would be appropriate if symmetry could not be assumed and comment on why the results differ.

2.6   A pathologist counts the numbers of diseased plants in randomly selected areas each 1 yard square on a large wheat field. For 35 such areas the numbers of diseased plants are:

21	17	43	81	32	102	117
43	39	11	67	23	142	7
44	39	82	93	28	145	0
17	77	53	50	60	9	14
40	19	101	104	33	2	22

Use an appropriate nonparametric test to find whether it is reasonable to assume the median number of diseased plants per square yard might be 50, (i) without assuming population symmetry, and (ii) assuming symmetry.

2.7   Before treatment with a new drug a number of insomniacs have a median sleeping time of 2 hours per night. It is known that the effect of the drug, if any, will be to increase sleeping time, but many doctors are doubtful whether it will have any effect at all. Is their doubt justified if the hours slept per night by the insomniacs after taking the drug are as follows?

3.1   1.8   2.7   2.4   2.9   0.2   3.7   5.1   8.3   2.1   2.4

Do not assume a symmetric distribution of sleeping times in the population.

2.8   In Section 2.2.3 we showed that for samples of 12 the critical value of the lesser of S_p, S_n for significance at the 5% level given by (2.4) was the same as that given by the exact theory. Is the situation the same at the 1% level?

2.9   The numbers of pages for all 114 statistics books on my shelves are given in Section A7.1. Select from these four independent random samples (see Section A3) each of 15 books. If θ is the population median number of pages, for each of the four samples use both the sign test and the Wilcoxon signed rank test to test $H_0: \theta = 230$ against $H_1: \theta \neq 230$. Comment on the implications of your results. For each sample obtain 95% confidence limits for θ both with and without an assumption of symmetry. Explain why the intervals differ and what these differences imply. Since we have all 114 page numbers recorded in Section A7.1 we can determine the true population value of θ. What is it (see Exercise 1.2)?

Is it included in the confidence intervals you obtain for all four samples? Would you expect it to be included in the 95% confidence interval based on **any** random sample of 15? If not, why not?

2.10 The Meteorological Office monthly weather summaries published by HMSO give the following rainfall in mm for 15 stations in England and Scotland during 1978. The stations are listed below in order of increasing longitude. Is there evidence of a monotonic trend in rainfall in 1978 as we move from South to North?

Margate, 443; Kew, 598; Cheltenham, 738; Cambridge, 556; Birmingham, 729; Cromer, 646; York, 654; Carlisle, 739; Newcastle, 742; Edinburgh, 699; Callender, 1596; Dundee, 867, Aberdeen, 877; Nairn, 642; Baltasound 1142.

2.11 In a small pilot opinion poll 18 voters in one electorate were selected at random and asked if they thought the British Prime Minister was doing a good job. Six (one-third) said 'yes' and 12 (two-thirds) said 'no'. Is this sufficient evidence to reject the hypothesis that 50% of the electorate think the Prime Minister is doing a good job? The pilot poll results were checked by taking a larger sample of 225 voters. By a remarkable coincidence 75 (one-third) answered 'yes' and 150 (two-thirds) answered 'no', exactly the same proportions as in the pilot survey. Do we draw the same conclusion about the hypothesis that 50% of the electorate think the Prime Minister is doing a good job? If not, why not?

2.12 A traffic warden is required to note the time a car has been illegally parked after its metered time has expired. For 16 offending cars he records the times in minutes as:

10 42 29 11 63 145 11 8 23 17 5 20 15 36 32 15

Obtain an appropriate nominal 95% confidence interval for the median overstay time of offenders prior to detection. What is the actual confidence level for the interval you obtain? What assumptions were you making to justify using the method you did, and to what population do you think the confidence interval might apply?

2.13 Kimura and Chikuni (1987) give data for lengths of Greenland turbot of various ages sampled from commercial catches in the Bering Sea and aged and measured by the Northwest and Alaska Fisheries Center. For 12-year-old turbot the numbers of each length were:

Length (cm)	64	65	66	67	68	69	70	71	72
No. of fish	1	2	1	1	4	3	4	5	3

Length (cm)	73	74	75	77	78	83
No. of fish	3	0	1	6	1	1

Would you agree with someone who asserted that, on this evidence, the median length of 12-year-old Greenland turbot was almost certainly between 69 and 72 cm?

2.14 The journal *Biometrics* (1985, **41**, 830) gives data on the numbers of medical papers published by that journal for the period 1971–81. These data are extended below to cover the period 1969–85. Is there evidence of a monotonic trend in numbers of medical papers published?

11   6   14   13   18   14   11   22   19   19   25   24   38   19   25   31   19

2.15 Knapp (1982) gives percentage of births on each day of the year averaged over 28 years for Monroe County, New York State. Ignoring leap years (which make little difference), the median percentage of births per day is 0.2746. Not surprisingly this is close to the expected percentage on the assumption that births are equally likely on any day, viz. $100/365 = 0.274$. We give below the average percentages for each day in the month of September. If births are equally likely to be on any day of the year this should resemble a random sample from a population with median 0.2746. Do the data confirm this?

0.277   0.277   0.295   0.286   0.271   0.265   0.274   0.274   0.278
0.290   0.295   0.276   0.273   0.289   0.308   0.301   0.302   0.293
0.296   0.288   0.305   0.283   0.309   0.299   0.287   0.309   0.294
0.288   0.298   0.289

2.16 A commentator on the 1987 Open Golf Championship at Muirfield, Scotland asserted that on a good day 10% of top-class players could be expected to complete a round with a score of 68 or less. On the fourth day of the championship weather conditions were poor and the commentator remarked before play started that we could expect the weather to increase scores by 4 strokes per round. This would suggest 10% of players might be expected to return scores of 72 or less. In the event 26 of the 77 players competing that day returned scores of 72 or less. Regarding the players as a sample of 77 top-class players and assuming the claim about the top 10% of scores on a good day is true, do these fourth-day results suggest that the commentator's assertion about scores under the poor weather conditions on the fourth day for top-class players was: (i) perhaps correct, (ii) almost certainly optimistic, or (iii) almost certainly pessimistic?

2.17 Rogerson (1987) gives the following annual mobility rates (percentage of population living in different houses at the end of the year than at the beginning) for people of all ages in the US for 28 consecutive post-war years. Is there evidence of a monotonic trend?

18.8   18.7   21.0   19.8   20.1   18.6   19.9   20.5   19.4   19.8
19.9   19.4   20.0   19.1   19.4   19.6   20.1   19.3   18.3   18.8
18.3   18.4   17.9   17.1   16.6   16.6   16.1   16.8

3

Distribution tests and rank transformations for single samples

3.1 MATCHING SAMPLES TO DISTRIBUTIONS

Chapter 2 covered test and estimation procedures for location – the median if we do not assume symmetry; either mean or median if we do. We now consider more general problems of matching data to particular distributions. Populations, whether or not they have identical locations, may differ widely in other characteristics.

We may want to know if observations are consistent with their being a sample from some specified continuous distribution. Kolmogorov (1933, 1941) devised a test for this purpose and Smirnov (1939, 1948) extended it to test whether two independent samples could reasonably be supposed to come from the same distribution. We consider here the Kolmogorov test.

Given observations $x_1, x_2, \ldots, x_n$ we may ask if the values are consistent with sampling from a completely specified distribution, e.g. a uniform distribution between 0 and 1 or between 0 and 10; or a normal distribution with mean 20 and standard deviation 2.7. Or we may ask a more general question like 'Is it reasonable to suppose these data are from some normal distribution, mean and variance unspecified?' Kolmogorov's test requires modification to answer the last question. We consider a possible modification in Section 3.1.3. The Kolmogorov test is distribution-free, because the same test procedure is used for any distribution completely specified under H_0.

3.1.1 Kolmogorov's test

The **continuous uniform distribution** (sometimes called the **rectangular distribution**) is probably the simplest of all continuous distributions. It is the distribution of a random variable which is equally likely to take a value in any fixed-length segment within a named interval. For example, suppose pieces of thread each exactly 6 cm in length are clamped at each end and a force applied until they break; if the thread always breaks at its weakest point and this is equally likely to be anywhere along its length, then the breaking point will be uniformly distributed over the interval (0, 6), where the distance to the break is measured in centimetres from the left-hand clamp.

The probability density function (see Section A1.1) has the form

$$f(x) = 0, x \leqslant 0; \qquad f(x) = 1/6, 0 < x \leqslant 6; \qquad f(x) = 0, x > 6.$$

The rectangular form of $f(x)$ gives rise to the name 'rectangular distribution'. If the thread always breaks the probability is 1 that it breaks somewhere in the interval 0 to 6. Thus the total area under the density curve must be 1; clearly only one form of density curve, that given above, satisfies both this condition and that of equal probability of a break in segments of the same length wherever they lie in the interval. The function is graphed in Figure 3.1 and is essentially a line running from 0 to 6 at height 1/6 above the x-axis. The lightly shaded area represents the total probability of 1 associated with the complete distribution. The heavily shaded rectangle PQRS between $x = 3.1$ and $x = 3.8$ represents the probability that the random variable X, the distance to the break, takes a value between 3.1 and 3.8. Note that it is conventional to use X, Y as names for random variables, and the corresponding lower case x, y (with or without subscripts) for specific values of these variables. Since PQ = $3.8 - 3.1 = 0.7$ and PS = 1/6, clearly this area is $0.7/6 = 0.1167$. This equals the probability of a break occurring in any segment of length 0.7 lying entirely within the interval (0, 6).

The Kolmogorov test does not use the probability density function, but the related cumulative distribution function, i.e. a curve giving $\Pr(X \leqslant x)$ for all x in (0, 6). In our example the probability that the break occurs in the first two

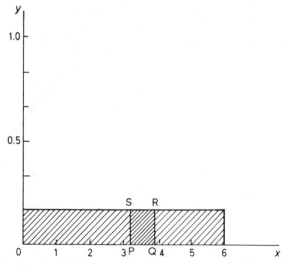

Figure 3.1 Probability density function for a continuous uniform distribution over (0, 6).

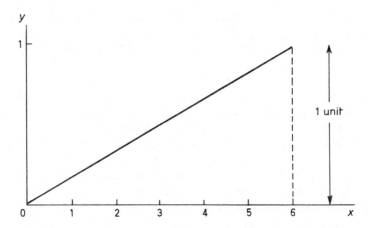

Figure 3.2 Cumulative distribution function for a uniform distribution over (0, 6).

centimetres is $\Pr(X \leqslant 2)$. Clearly this is $1/3$. The cumulative distribution function is written $F(x)$ and for any x between 0 and 6, $F(x) = x/6$. It has the value 0 at $x = 0$ and 1 at $x = 6$. It is graphed in Figure 3.2.

These notions generalize to a uniform distribution over any interval (a, b) and the cumulative distribution function is then a straight line rising from zero at $x = a$ to 1 at $x = b$, representing $F(x) = (x - a)/(b - a)$, $a \leqslant x \leqslant b$. Clearly $F(a) = 0$ and $F(b) = 1$; this is illustrated in Figure 3.3.

For other distributions the cumulative distribution functions are less simple. However, for continuous distributions they are always smooth curves starting at zero, moving upward from left to right, and finally reaching the value 1. They may be flat in parts but never decrease, and are said to be monotonic non-decreasing (or monotonic increasing). Figure 3.4 shows the form of the cumulative distribution function for the standard normal distribution, i.e. the

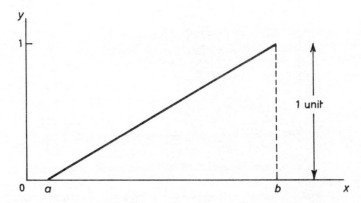

Figure 3.3 Cumulative distribution function for a uniform distribution over (a, b).

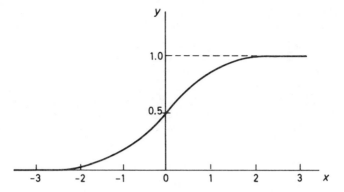

Figure 3.4 Cumulative distribution function for the standard normal distribution

normal distribution with zero mean and standard deviation 1.

In essence the Kolmogorov test compares a population cumulative distribution function with a related curve $S(x)$ based on sample values and called the **sample (or empirical) cumulative distribution function**. For a sample of n observations

$$S(x) = \frac{\text{number of sample values less than or equal to } x}{n} \tag{3.1}$$

Example 3.1

The problem. Below are the distances from one end at which each of 20 threads each 6 cm long break when subjected to strain. Form and graph $S(x)$. For convenience the distances are ordered.

0.6 0.8 1.1 1.2 1.4 1.7 1.8 1.9 2.2 2.4
2.5 2.9 3.1 3.4 3.4 3.9 4.4 4.9 5.2 5.9

Formulation and assumptions. From (3.1) it is clear that $S(x)$ increases by $1/20$ at each unique x value corresponding to a break, or by $r/20$ if r break distances coincide.

Procedure. When $x = 0$, $S(x) = 0$; it keeps this value until $x = 0.6$, the first break. Then $S(0.6) = 1/20$, and $S(x)$ maintains the value $1/20$ until $x = 0.8$ when it jumps to $2/20$, a value retained until $x = 1.1$. It jumps in steps of $1/20$ at each break value until $x = 3.4$, where it increases by $2/20$ since two breaks occur at $x = 3.4$.

$S(x)$ is referred to as a step function for obvious reasons. Its value at each step is given in Table 3.1.

Conclusion. For this data $S(x)$ takes the form in Figure 3.5.

Table 3.1 Values of $S(x)$ at step points

x	0.6	0.8	1.1	1.2	1.4	1.7	1.8	1.9	2.2	2.4
$S(x)$	1/20	2/20	3/20	4/20	5/20	6/20	7/20	8/20	9/20	10/20

x	2.5	2.9	3.1	3.4*	3.9	4.4	4.9	5.2	5.9
$S(x)$	11/20	12/20	13/20	15/20	16/20	17/20	18/20	19/20	1

* Repeated sample value.

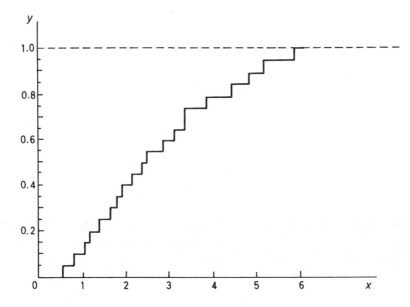

Figure 3.5 Sample cumulative distribution function for thread breaks.

Comment. $S(x)$ is a sample estimator of the population cumulative distribution function $F(x)$. If a sample really comes from a specified distribution, the step function $S(x)$ should never depart markedly from the population cumulative distribution function $F(x)$. In this example, if the breaks are uniformly distributed over $(0,6)$, one expects, for example, about half of them to be in the interval $(0, 3)$ so that $S(3)$ should not have a value very different from 0.5, and so on.

The idea that $S(x)$ should never depart too violently from $F(x)$ if our sample comes from a distribution with this cumulative distribution function is fundamental to Kolmogorov's test. The test is based on the maximum difference in magnitude between $F(x)$ and $S(x)$.

Example 3.2

The problem. Given the 20 breaking strengths and their $S(x)$ values in Table 3.1, is it reasonable to suppose breaking strengths are uniformly distributed over (0, 6)?

Formulation and assumptions. We seek the maximum difference in magnitude between $F(x)$ and $S(x)$; this is compared with a tabulated value to see if significance is indicated.

Procedure. Table 3.2 shows the values of $F(x)$ and $S(x)$ at each break point given in Table 3.1. It also gives at each of these points the difference $F(x_i) - S(x_i)$. We cannot guarantee that the maximum difference will occur among this set. Figure 3.6 shows what may happen.

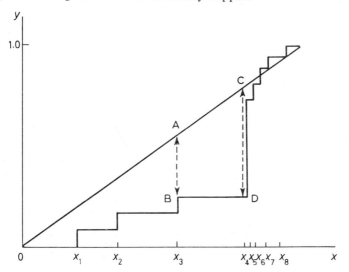

Figure 3.6 Locating a maximum difference between $F(x)$ and $S(x)$.

In this figure, for simplicity, we examine only eight distinct steps – some of which may be multiple steps – in $S(x)$, and suppose $F(x)$ is a straight line. Clearly the largest value of $F(x_i) - S(x_i)$ is at $x = x_3$, and is represented by AB, yet the greatest distance between the two functions occurs just below x_4 and is represented by CD; clearly its value is $F(x_4) - S(x_3)$. For this reason we record the differences $F(x_i) - S(x_{i-1})$ also in Table 3.2.

Figure 3.7 shows the graphs of $F(x)$ and $S(x)$. For much of its course $S(x)$ lies above $F(x)$. The entry of greatest magnitude in the last two columns of Table 3.2 is -0.18 when $x_i = 3.4$. Table A3 tells us that when $n = 20$ the largest difference (whether positive or negative) must be at least 0.294 for significance at the 5% level.

Table 3.2 Comparison of $F(x)$ and $S(x)$ for thread breaks

x_i	$F(x_i)$	$S(x_i)$	$F(x_i) - S(x_i)$	$F(x_i) - S(x_{i-1})$
0.6	0.10	0.05	0.05	0.10
0.8	0.13	0.10	0.03	0.08
1.1	0.18	0.15	0.03	0.08
1.2	0.20	0.20	0.00	0.05
1.4	0.23	0.25	-0.02	0.03
1.7	0.28	0.30	-0.02	0.03
1.8	0.30	0.35	-0.05	0.00
1.9	0.32	0.40	-0.08	-0.03
2.2	0.37	0.45	-0.08	-0.03
2.4	0.40	0.50	-0.10	-0.05
2.5	0.42	0.55	-0.13	-0.08
2.9	0.48	0.60	-0.12	-0.07
3.1	0.52	0.65	-0.13	-0.08
3.4	0.57	0.75	-0.18	-0.08
3.9	0.65	0.80	-0.15	-0.10
4.4	0.73	0.85	-0.12	-0.07
4.9	0.82	0.90	-0.08	-0.03
5.2	0.87	0.95	-0.08	-0.03
5.9	0.98	1.00	-0.02	0.03

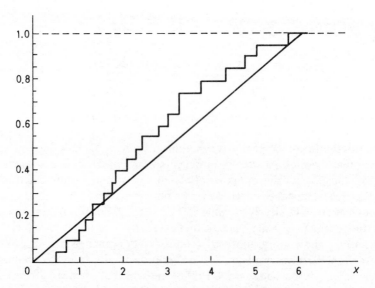

Figure 3.7 Graph of $F(x)$ and $S(x)$ for thread break data.

Conclusion. Since the largest difference has magnitude only 0.18 we have insufficient evidence at the 5% level to reject the null hypothesis that breaks are uniformly distributed.

Comments. A two-tail test is appropriate if our alternative hypothesis is that the observations may come from any other unspecified distribution. Practical situations may occur when the only alternative is one in which the cumulative distribution function must lie either only above or only below that specified by the null hypothesis. For example, in the thread problem it may be known that the test machine places stresses on the string that, if not uniformly distributed, will be higher at the left end of the string and decrease steadily as we move along it. This would increase the tendency for breaks to occur close to the left end of the string. With a high probability of breaks towards the left, decreasing as we move along the string, it is clear that the cumulative density would everywhere lie above the straight line representing $F(x)$ for the uniform distribution. Thus a maximum difference with $S(x)$ greater than $F(x)$ for the uniform distribution would, if significant, favour the acceptable alternative. Thus for a one-sided alternative hypothesis we use a one-tail test and only reject the null hypothesis if the largest difference has the appropriate sign.

In our thread example if our one-sided alternative were that the population cumulative distribution function was always at or above that for the uniform distribution, a negative difference would be appropriate in Table 3.2. Table A3 indicates that when $n = 20$ the magnitude must then exceed 0.265 for significance at the 5% level.

Is the Kolmogorov test wasting information by using only the difference of greatest magnitude? Tests have been proposed that take account of all differences, but these tend to have little, if any, advantages. This is not as contrary to intuition as it may at first seem, for the value of $S(x)$ depends at any stage on how many observations are less than the current value of x; therefore we are making the comparison at each stage on the basis of accumulated evidence to that stage.

The theory of the test is beyond the scope of this book. If the test is applied to a discrete rather than a continuous distribution it will tend to give too few rather than too many significant results and in this sense is described as conservative. We consider other tests for discrete distributions in Chapter 8.

3.1.2 A confidence region for a cumulative distribution function

We may determine a region with a specified degree of confidence – say 95% – that $F(x)$ lies within that region, by working from Kolmogorov significance test arguments.

Example 3.3

The problem. Determine a 95% confidence region for the true cumulative distribution function $F(x)$ for the thread break data of Example 3.2.

Formulation and assumptions. Given $S(x)$ for the data, we seek a region about $S(x)$ for which in the usual confidence level terminology we can be 95% confident that it contains the true but now supposedly unknown $F(x)$ entirely within its boundaries. The boundaries are determined by placing them at appropriate distances vertically above and below $S(x)$ subject only to the constraints that the boundaries do not allow $F(x)$ outside the range 0 to 1, since a cumulative distribution function cannot take values outside that range.

Procedure. The values of $S(x)$ at the step points for this example are given in Tables 3.1 and 3.2 and the function is graphed by the dark step function in Figure 3.8. We noted in Example 3.2 (from Table A3) that the minimum value for significance with a two-tail test at the 5% level when $n = 20$ was $\max|F(x) - S(x)| = 0.294$. The two lighter step functions in Figure 3.8 are at vertical distances 0.294 above and below $S(x)$ subject to the constraints that they must lie between 0 and 1. For any $F(x)$ lying entirely within the region between these outer step functions we would accept the hypothesis that the sample came from such a population, as then $|F(x) - S(x)|$ would never exceed 0.294.

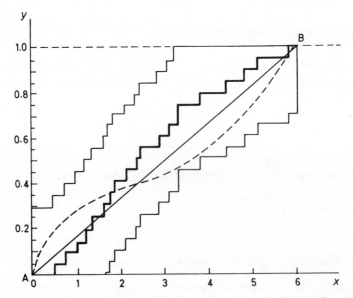

Figure 3.8   Confidence region for a population cumulative distribution function.

Conclusion. The region enclosed by the outer step functions in Figure 3.8 is a 95% confidence region for $F(x)$.

Comment. The straight line AB in Figure 3.8 is the cumulative uniform (0, 6) distribution function. The dotted curve in that figure indicates another form of $F(x)$ that would (just) be acceptable; this is simply a freehand curve with properties of a cumulative distribution function; it is not based on any mathematical model involving parameters and thus shows well the truly nonparametric or distribution-free nature of our procedure for obtaining a confidence region.

3.1.3 A test for normality

With a sample from a normal distribution, optimum test and estimation procedures are based on the appropriate normal theory; before using these we may want to test whether a sample may reasonably be supposed to come from a normal distribution. The Kolmogorov test is appropriate to decide whether a sample may come from a normal distribution with known parameters μ, σ^2. This is no longer the case if we have to estimate the parameters from the data; the same statistic may still be used but the critical values in Table A3 are no longer appropriate.

This is an important practical situation and a test proposed by Lilliefors (1967) is useful. We assume our observations are a random sample from some unspecified continuous distribution, and test whether it is reasonable to suppose this a member of the normal family. In this sense our test is not distribution-free.

The basic ideas can be extended to test compatibility with other sets of distributions such as the gamma family with unspecified parameter values, but separate tables of critical values are required for each family of distributions. An interesting discussion of the difficulties in obtaining critical values is given in Conover (1980, Section 6.2). Conover (Section 6.2) also discusses the Shapiro–Wilk test for normality which again requires special tables. This test has good power against a wide range of alternatives but the rationale is not so easily described by intuitive arguments. Bowman and Shenton (1975) describe another class of test for normality based on estimates of skewness and kurtosis which involve calculation of the first four sample moments. These are intuitively appealing and one of their tests is shown by Jarque and Bera (1987) to be powerful against a wide range of alternatives. However, departures from normality do not always reflect themselves in marked changes in skewness and kurtosis, especially if there is bimodality.

These tests for normality are not strictly nonparametric, so we describe only the Lilliefors test and demonstrate its similarity to the Kolmogorov test.

Critical values for the Lilliefors test for normality for rejection at the 5% and 1% levels are given in Table A4.

Example 3.4

The problem. In the Badenscallie burial ground in Wester Ross, Scotland, the ages at death of males were noted on all tombstones for four prominent clans in that district. From all 117 ages recorded a random sample of 30 was taken and the ages for that sample were, in order,

$$11 \quad 13 \quad 14 \quad 22 \quad 29 \quad 30 \quad 41 \quad 41 \quad 52 \quad 55 \quad 56 \quad 59 \quad 65 \quad 65 \quad 66$$
$$74 \quad 74 \quad 75 \quad 77 \quad 81 \quad 82 \quad 82 \quad 82 \quad 82 \quad 83 \quad 85 \quad 85 \quad 87 \quad 87 \quad 88$$

Is it reasonable to suppose the death ages are normally distributed?

Formulation and assumptions. We use the Lilliefors test. This is carried out using the Kolmogorov statistic to compare $S(x)$ and the standard normal variable cumulative distribution function after standardizing the data by the transformation

$$z_i = \frac{x_i - m}{s}$$

where m is the sample mean and s the usual estimate of population standard deviation, obtained from

$$s^2 = \left[\sum_i (x_i^2) - \left(\sum_i x_i \right)^2 \Big/ n \right] \Big/ (n - 1)$$

Procedure. We find $m = 61.43$ and $s = 25.04$. Successive z_i are calculated; e.g. for $x_i = 11$, $z_i = (11 - 61.43)/25.04 = -2.014$. Tables for the standard normal variable (see, e.g. Neave, 1981, pp. 18–19) show $F(-2.014) = 0.022$.

Denoting the standard normal distribution function by $F(z)$ and the sample cumulative distribution function by $S(z)$ we set up Table 3.3 in analogous manner to Table 3.2. Although not essential, an additional column giving the x-values has been included; an asterisk against a value implies it occurs more than once in the data.

The largest difference is 0.192 occurring in the final column and corresponding to the sample values 74. From Table A4 we see that the 5% and 1% critical values when $n = 30$ are respectively 0.161 and 0.187.

Conclusion. A maximum difference between step function and distribution function of 0.192 is highly significant, so we reject at the 1% level the null hypothesis that the data are from any normal population.

Comments. 1. A glance at the sample data suggests that a few of the males have died young, and that the distribution is somewhat skew; a large number

Table 3.3   The Lilliefors normality test. Badenscallie ages at death

x	z	$F(z)$	$S(z)$	$F(z_1) - S(z_i)$	$F(z_i) - S(z_{i-1})$
11	-2.014	0.022	0.033	-0.011	0.022
13	-1.934	0.026	0.067	-0.044	-0.007
14	-1.894	0.029	0.100	-0.071	-0.038
22	-1.575	0.058	0.133	-0.075	-0.042
29	-1.295	0.098	0.167	-0.069	-0.035
30	-1.255	0.105	0.200	-0.095	-0.062
41*	-0.816	0.207	0.267	-0.060	-0.007
52	-0.377	0.353	0.300	0.053	0.086
55	-0.257	0.399	0.333	0.066	0.099
56	-0.217	0.414	0.367	0.047	0.081
59	-0.097	0.461	0.400	0.061	0.094
65*	0.142	0.556	0.467	0.089	0.156
66	0.183	0.572	0.500	0.072	0.105
74*	0.502	0.692	0.567	0.125	0.192
75	0.542	0.706	0.600	0.106	0.139
77	0.622	0.733	0.633	0.100	0.133
81	0.781	0.782	0.667	0.115	0.149
82*	0.821	0.794	0.800	-0.006	0.127
83	0.861	0.805	0.833	-0.028	-0.005
85*	0.942	0.827	0.900	-0.073	-0.006
87*	1.021	0.846	0.967	-0.121	-0.054
88	1.061	0.856	1.000	-0.144	-0.111

of deaths occur after age 80. Figure 3.9 shows the sample data on a histogram with class interval 10. Readers familiar with typical histograms for samples from a normal distribution will not be surprised that we reject the hypothesis of normality. Indeed much 'life-span' data, whether it be for men, animals or even for the number of days machines function without a breakdown, tend characteristically to have a non-normal distribution.

2. Full data on ages at death for all four prominent clans recorded from tombstones in the Badenscallie burial ground are given in Section A7.2.

3.2 ROBUSTNESS

In recent years there has been increasing interest in nonparametric methods because the assumptions needed for their validity are so broad that they apply to data that vary widely in their characteristics. This applies particularly to field or laboratory data that are collected in surveys; less perhaps to controlled

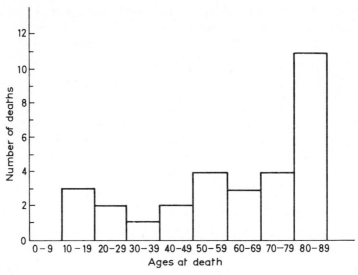

Figure 3.9 Histogram of death ages at Badenscallie.

experimental data. In all these situations there is often the odd observation that looks peculiar, but there is no valid reason to reject it; it may or may not be a sensible and correct result. For this reason interest has focused on methods that are not greatly affected by 'peculiar' observations, i.e. that give similar results whether or not that observation is genuine. These are the class of **robust** methods. They are of added importance now that much data is 'untouched by human hands' and processed by computer. If an observation is recorded as 332 when it should be 33 or 32 this may go undetected. (How easy it is to touch the 3 and 2 keys on a keyboard almost simultaneously and get an entry 32 or 23 for a 3 or 2!). The careful data analyst tries to avoid such mistakes; their effect on a simple *t*-test may be devastating as the following example shows.

Example 3.5

The problem. An experimenter wants 95% confidence limits for the mean based on these data:

$$7 \quad 11 \quad 5 \quad 14 \quad 10 \quad 8 \quad 12 \quad 13 \quad 19$$

Unfortunately the data are transmitted as

$$7 \quad 11 \quad 5 \quad 14 \quad 10 \quad 8 \quad 12 \quad 113 \quad 19$$

How does this affect confidence limits if they are based on: (i) the normal theory *t*-distribution; (ii) the sign test; (iii) Wilcoxon signed ranks?

Formulation and assumptions. The confidence limits are calculated for each set of data by each of the three methods and the results compared.

Procedure. The 95% confidence intervals based on the normal theory *t*-distribution are calculated using the method in Section A4.1. For the two data sets they turn out to be respectively (5.8, 12.2) and (-3.8, 48.0).

The observations in ascending order are

$$5 \quad 7 \quad 8 \quad 10 \quad 11 \quad 12 \quad 13 \quad 14 \quad 19$$

and

$$5 \quad 7 \quad 8 \quad 10 \quad 11 \quad 12 \quad 14 \quad 19 \quad 113$$

Nominal 95% (actual 96.1%) confidence intervals for the sign test calculated in the usual way are (7, 14) and (7, 19) respectively.

Finally, if we use the Wilcoxon rank sums, we see from Table A2 that we reject the 5 smallest and 5 largest 'paired' means in the sense described in Section 2.2.6. It is easily seen in the first case that the six smallest paired means are 5, 6, 6.5, 7, 7.5, 7.5 and the six largest are 19, 16.5, 16, 15.5, 15, 14.5 giving (7.5, 14.5) as the confidence interval. In the second case the six smallest values are the same as those above but the six largest are 113, 68, 63.5, 62.5, 62, 61.5 giving (7.5, 61.5) as the confidence interval.

Conclusion. Confidence intervals based on the sign test are affected minimally by the gross error. This is not surprising as we do not need an assumption of symmetry for validity of a sign test. By any reasonable criteria the erroneous data set are asymmetric! Both the normal theory and Wilcoxon confidence intervals are greatly affected by the rogue 119; as both assume symmetry this is not surprising.

Comments. 1. While neither normal theory nor Wilcoxon limits are appropriate with the obvious asymmetry of the second data set, at least the Wilcoxon limits lie within the data range. The confidence interval based on normal theory gives limits outside the data range. Applying a *t*-test to the given data, testing at the 5% level we would accept the hypothesis that the data mean may be zero (a value outside the data range).

2. It is easy to scoff at those using normal theory tests or tests that assume symmetry for the second and erroneous data set; but such things are sometimes done by experienced research workers who are not ignorant of basic statistics. The increasing tendency to leave analysis of uninspected data to a statistical computer package enhances this danger. This is not to advocate that we should not use statistical packages, but until they are developed with 'expert system' ability to detect inappropriate data, one must emphasize the cardinal rule of looking at data before analysis, or if this is impossible because

data processing is highly automated, at least every result must be scrutinized to see that it does not fly in the face of reason.

3. The Lilliefors test rejects normality of the second data set. See Exercise 3.4.

We look at robust methods in a more general context in Chapter 9.

3.3 TRANSFORMATIONS OF RANKS

As it is based on ranks, the Wilcoxon signed rank test is not using full information from measurement data. One may reasonably ask if there are other transformations that preserve rank order and yet allow us to produce tests analogous to and perhaps superior to the Wilcoxon test. The simplest 1:1 transformation based on ranks is the data arranged in ascending order. Although we usually use ordered data to obtain ranks it is also valid to look at the ordered data itself as a 1:1 transformation of ranks. We may carry out a randomization test analogous to the Wilcoxon test on such ordered data. We illustrate the procedure by a simple example, but as we shall see, the method has serious practical limitations although it nearly always gives results similar to the signed rank test when this is valid; i.e. if symmetry of the population distribution can be assumed. The approach is of historical interest as it was developed before the Wilcoxon test with the aim of justifying normal distribution theory in most cases where a symmetry assumption was valid. Pioneers in this area were Fisher (1935) and Pitman (1937, 1938), who first developed randomization tests.

Example 3.6

The problem. Given the observations 98, 107, 112, 93, 149, 85, 122 from a distribution assumed symmetric with unknown mean μ, test the hypothesis $H_0: \mu = 95$ against $H_1: \mu > 95$, basing the test on the ordered observations.

Formulation and assumptions. Since symmetry is assumed it follows that not only is there an equal probability of any observation being above or below the mean, but probabilities of deviations of any given magnitude are equally likely; e.g. we are just as likely to have a sample value between 3 and 4 units above the mean as we are to have one between 3 and 4 units below. Thus samples that have a comparable amount of scatter above and below the hypothetical mean support the null hypothesis, whereas samples with more and larger deviations in one direction than in the other provide evidence against the null hypothesis. We show how to assess this evidence in the following procedure section. We have a one-sided H_1, so a one-tail test is appropriate.

Procedure. Under H_0 for our data the deviations from the mean, 95, are respectively 3, 12, 17, -2, 54, -10, 27. We may set these out formally:

Negative deviations: 2, 10
Positive deviations: 3, 12, 17, 27, 54.

There are two negative and five positive deviations; four of the positive deviations are larger in magnitude than either negative deviation. Intuitively this makes us a little suspicious about the population mean being 95. We ask formally: given the magnitudes of deviations listed above, and considering all possible allocations of positive and negative signs attached to them, which allocations fall into a critical region for which we reject H_0? That is, we now do with the actual deviations what we did with the ranks in the Wilcoxon test.

The strongest indication against the null hypothesis, and in favour of H_1, would have been if all deviations from 95 had been positive. As in the Wilcoxon test, the number of ways we can allocate signs (in the Wilcoxon case to seven ranks, here to seven observed differences) is $2^7 = 128$.

If we denote by S_p and S_n the sum of positive and negative deviations respectively, we take groupings with low values of S_n (and consequently high values of S_p) to indicate significance. Since $S_p + S_n = S$, the (fixed) sum of all absolute deviations, we may concentrate on low S_n values only.

Clearly the lowest negative sum occurs when there are no observations in the group (i.e. all deviations positive); then $S_n = 0$. It is easy to see the groupings with next lowest sums are 2 only (sum 2), 3 only (sum 3), 2 and 3 (sum 5), 10 only (sum 10) then 12 or 2 and 10 (both with sum 12). There are thus seven groupings that give sums less than or equal to 12 for negative signs. We conclude:

$$\Pr(S_n \leqslant 12) = 7/128 = 0.055.$$

Similarly, it is easily seen that

$$\Pr(S_n \leqslant 10) = 5/128 = 0.039.$$

This implies that for significance at the 5% level we require $S_n \leqslant 10$ in the relevant one-tail test.

Conclusion. Since, in our example, $S_n = 12$ we cannot reject H_0.

Comments. 1. Modifications for a two-tail test are straightforward; see Exercise 3.5.

2. For even moderate numbers of observations it is tedious to list all cases in the critical region. The process is also error prone; one has to start afresh with each problem and list all relevant sets. The approach has a strong intellectual appeal as it uses all relevant information in the data, but it is not very practical.

In their pioneer work, Fisher and Pitman showed that under a wide range of conditions the results of randomization tests and certain parametric tests were very similar and their arguments were in some ways a justification of parametric tests when some assumptions were only approximately fulfilled.

Modern developments have indicated that there is little loss in efficiency by replacing the deviations by their ranks as in the Wilcoxon test. As the relevant value of the signed rank sum for significance then depends only on the sample size n, this enables us to use Table A2 (or relevant large-sample normal approximations) for testing and estimation with any sample of size n from any symmetric distribution.

3.3.1 Normal scores

Transformation of continuous ordered data to ranks is in essence a transformation to something like a uniform distribution – the values being in a sense idealized values placed at the quantiles of such a distribution. A further transformation of ranks gives values at corresponding quantiles of a normal distribution. There are several modifications discussed by Conover (1980, pp. 316–27) and others. If for n ranked observations we replace rank i by the $i/(n+1)$ quantile ($i = 1, 2, \ldots, n$) of a standard normal distribution, we get a set of observations with mean zero. For example, if our original observations are 2, 3, 7, 21, 132 these are very non-normal. The ranks are 1, 2, 3, 4, 5 – an idealized sample from a uniform distribution. The normal scores would be the 1/6th, 2/6th, 3/6th, 4/6th, and 5/6th quantiles of the normal distribution. These may be read from cumulative normal distribution tables. The normal score corresponding to $1/6 = 0.1667$ is the x-value such that the standard normal cumulative distribution function $F(x) = 0.1667$. From tables (e.g. Neave, 1981, p. 18) we find $x = -0.97$. That corresponding to $2/6 = 0.3333$ is -0.43, and that corresponding to $3/6 = 0.5$ (the mean or median) is clearly zero; corresponding to $4/6 = 0.6667$ it is 0.43 and corresponding to $5/6 = 0.8333$ it is 0.97. Note the symmetry. These quantile scores are sometimes referred to as **van der Waerden scores**, their use being proposed by B. L. van der Waerden.

An alternative to van der Waerden scores are **expected normal scores** where the ith ordered sample value is replaced by the expectation of the ith order statistic for a normal distribution. The ith order statistic is the ith smallest value in a sample of n observations. If the observations are arranged in ascending order we often write the ith order statistic as $x_{(i)}$. Fisher and Yates (1957, Table XX) give expected normal scores corresponding to ranks in samples of various sizes. In practice van der Waerden scores and expected normal scores usually lead to almost identical conclusions and we illustrate the use of the former only.

A complication arises that precludes direct application of normal score transformations as an alternative to the Wilcoxon signed rank method; we essentially consider allocation of signs to the magnitudes of ranks in that method. The difficulty may be overcome if we have a sample of n observations by taking scores corresponding only to the n positive scores that would be allocated to a sample of $2n + 1$ observations; i.e. corresponding to the $(n + 2)/(2n + 2)$th, $(n + 3)/(2n + 2)$th, ..., $(2n + 1)/(2n + 2)$th quantiles. We then assign negative signs to any of these scores for which the corresponding rank is negative. Another way of expressing the procedure is to say that we get scores corresponding to the $\frac{1}{2}[1 + r/(n + 1)]$ quantiles for $r = 1, 2, ..., n$. For the rank magnitudes 1, 2, 3, 4, 5 the scores now become 0.21, 0.43, 0.67, 0.97, 1.38. Unlike the van der Waerden scores, these are no longer symmetric, but they are nevertheless more like a sample from a normal (but not standard normal) distribution than are the ranks themselves.

We devise a test similar to the Wilcoxon signed rank test but based on these scores. We allocate signs to scores as we did to ranks and then obtain the sum, W, of all signed scores s_i. An exact theory exists, but because we start with a near normal pattern of observations, approximate normal theory tests may be used even for quite small samples with or without ties. The test statistic

$$T = W \bigg/ \left\{ \sqrt{\left(\sum_i s_i^2 \right)} \right\}$$

has an approximately standard normal distribution under the null hypothesis of equal means. This follows because the mean of the score sum is zero (balance of positive and negative) under the null hypothesis. The sum of all signed scores is then approximately normally distributed with mean 0 and variance the sum of squares of scores, since the correction for the mean is zero. T is the analogue of the alternative Wilcoxon signed rank statistic, Z, introduced in (2.6), Section 2.2.5.

Example 3.7

The problem. Use a normal score test to test $H_0: \theta = 220$ against $H_1: \theta \neq 220$ for the data on numbers of pages used in Example 2.6. The data are

126   142   156   228   245   246   370   419   433   454   478   503

Formulation and assumptions. We replace signed ranks by the modified scores for the $\frac{1}{2}[1 + r/(n + 1)]$th quantiles where $n = 12$.

Procedure. The signed ranks of deviations (see Example 2.6) are 1, 2, 3, -4, -5, -6, 7, 8, 9, 10, 11, 12. Ignoring temporarily the signs, we first transform each to $\frac{1}{2}[1 + r/13]$ and then look up the corresponding quantile in tables of the standard normal distribution, e.g. for $r = 1$ we require the ordinate z_1 such that $\Pr(Z < z_1) = \frac{1}{2}(1 + 1/13) = 7/13 = 0.5385$, whence, from tables, $z_1 = 0.10$. It suffices to work with scores to two decimal places. Proceeding this way we get the following scores:

Rank	1	2	3	4	5	6	7	8	9	10	11	12
Score	0.10	0.19	0.29	0.40	0.50	0.62	0.74	0.87	1.02	1.20	1.43	1.77

We now restore signs to scores corresponding to those in the original ranks giving $0.10, 0.19, 0.29, -0.40, -0.50, -0.62, 0.74, 0.87, 1.02, 1.20, 1.43, 1.77$. The sum of these is 6.09 and the sum of squares is 9.8873. Our test statistic, $T = 6.09/\sqrt{9.8873} = 1.94$.

Conclusion. The value 1.94 is just short of the value 1.96 required for significance in a two-tail test, consistent with our finding for the Wilcoxon test on ranks, but the result is here very close to the critical value. One would usually report it as almost reaching significance at the 5% level.

Comments. 1. Normal score tests in this and other situations nearly all have the remarkable asymptotic property that their Pitman efficiency is 1 when the alternative normal theory test is appropriate and in any other case the Pitman efficiency is equal to or greater than 1. Euphoria so generated must be tempered with the realization that there are other tests with Pitman efficiency greater than 1 relative to normal theory tests for certain appropriate distributional assumptions, and also that an asymptotic result does not guarantee equivalent small-sample efficiency, although Pitman efficiency is usually a good guide.
2. By considering other functions of ranks, a whole range of tests may be devised, but in practice those based on normal scores or some simple modification to give only positive scores, such as the one used above, have most to commend them.

3.4 PRACTICAL IMPLICATIONS OF EFFICIENCY

It is clearly intuitively reasonable that the more relevant information we use, the better a testing or estimation procedure should be in terms of power and efficiency; it is now well established that in most cases where Pitman efficiency is not too low in the worst possible circumstances, yet high in more favourable ones, then, if there is any doubt about distributional assumptions, we should not lose much efficiency, if any, by using the most appropriate nonparametric test. For the Wilcoxon signed rank test compared with the t-test the Pitman efficiency is $3/\pi = 0.955$, when the t-test is optimal. This rises to 1 when the sample data are from a uniform distribution, and to 1.5 if the data are a sample from the sharply peaked and long-tailed double exponential distribution. In no circumstances is the Pitman efficiency of the Wilcoxon test relative to the *t*-test less than 0.864.

We now compare confidence intervals obtained by several methods using a fairly large body of data.

Example 3.8

The problem. For one particular clan – it would be invidious to give the real name, so we shall call them the McAlpha clan (but the data are real) –ages at death were recorded for all males of that clan buried in the Badenscallie burial ground.

For all 59 burials the ages (arranged for convenience in ascending order) were:

0	0	1	2	3	9	14	22	23	29	33	41	41	42	44
52	56	57	58	58	60	62	63	64	65	69	72	72	73	74
74	75	75	75	77	77	78	78	79	79	80	81	81	81	81
82	82	83	84	84	85	86	87	87	88	90	92	93	95	

Obtain and compare nominal 95% confidence limits for the population median using: (i) the sign test procedure, (ii) the Wilcoxon signed rank procedure, and (iii) normal distribution theory.

Formulation and assumptions. We use the standard methods already given, using large-sample approximations as appropriate. In the comments below we examine the validity of assumptions in each case.

Procedure. Using the normal approximation to the binomial distribution with $n = 59$, $p = \frac{1}{2}$, and the methods for confidence intervals given in Section 2.1.2 for the sign test approach, a 95% confidence interval is (62, 78). The Wilcoxon procedure given in Section 2.2.6 is very time consuming without a computer as one must eliminate the 624 largest and 624 smallest mean-pair estimates. The resulting confidence interval is (56.75, 73.5). The normal theory confidence interval is (54.7, 68.9).

Conclusion. The intervals are positioned somewhat differently; that based on normal assumptions is a little shorter than the other two, which are of similar width. Some care is needed in interpreting these results.

Comments. 1. The sign-test interval is essentially an interval for a median; the other two (assuming symmetry) can be regarded as intervals for a mean or median as these coincide under a population symmetry assumption (needed for validity of the methods). What about the normality assumption? The Lilliefors test (tedious without a computer program for a sample this size) indicates an assumption of normality is not justified. The test statistic, the maximum difference between the standardized step function and the standard normal density function is, for this data, 0.203 compared to a critical value of 0.134 for significance at the 1% level. Indeed, had we an available critical value for the 0.1% level it is likely this would also be exceeded. There is strong evidence for rejecting the normality assumption. Figure 3.10 is a histogram with class interval 10 for these data.

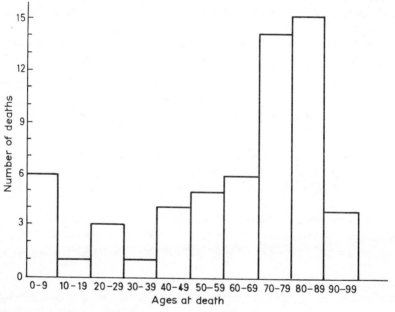

Figure 3.10 Histogram of death ages for clan McAlpha.

Not only does the histogram indicate non-normality but indeed bi-modality – a small peak of infant deaths; few deaths in early adulthood; and then a build-up to many deaths in old age. This very skewness invalidates a basic assumption of the Wilcoxon signed rank test. The sign test, on the other hand, tells us something valid about the median of a skew distribution. The fact that the other tests lower the interval and bring it nearer the sample mean than the sample median (the sample mean is 61.8, the median 74) indicates the sort of compromise our tests are being forced to make to accommodate an incompatibility between estimates and assumptions. Of the techniques illustrated, that for the sign test is the only valid one and it implies something about the median. The normal theory test tells us something about the mean – the confidence interval is centred at the sample mean but it is not a relevant interval because of the failure of the normality assumption. The confidence level may not even be near 95% and certainly has no relevance to the median.

2. What is the relevance to reality of any estimation procedure in a case like this? From what population is this a random sample? It is clearly not a random sample of all buried in the Badenscallie burial ground, for families have different lifespan characteristics. Perhaps the pattern of death ages is a reasonable approximation to that for McAlphas in Wester Ross, Sutherland or the Western Isles (where the clan McAlpha is well established). The main use of these data is for comparisons with patterns for other clans, something we consider in Section 5.1.5.

3. Difficulties like those described in this example arise with data collection generally and are not confined to nonparametric tests. The differences illustrated when we analyse the data with various assumptions draw our attention to such problems.

3.5 MODIFIED ASSUMPTIONS

We have considered tests assuming our observations are n independent random observations all from the same population. Appropriate location tests in this and the previous chapter may be applied to n samples each of one observation from n different populations each having the same unknown median or mean, but differing in other distributional characteristics. It is hard to visualize many such situations in practice, but they may arise with biochemical or mechanical measurements; e.g. oil delivery from different machines may be known under certain circumstances to have symmetric distributions with a certain mean but variation about this mean may differ from machine to machine. A question that may then arise is: does the mean delivery alter if we change the operating temperature? Testing this would be appropriate if there were physical reasons for believing temperature change would alter, if anything, the mean, and that it would do so by the same amount for all machines.

3.6 FIELDS OF APPLICATION

Tests for location in this chapter are relevant to many of the examples given in Section 2.4. For some of those examples, more general tests about distributions of the Kolmogorov type, or the Lilliefors test for normality, may also be relevant.

The Kolmogorov test may be used with any random sample providing we have some specific continuous population distribution in mind. If we wish to test whether our sample might reasonably be supposed to come from a normal population (mean and variance unspecified) the Lilliefors, or some other test for normality, is appropriate. Some examples follow.

Biology

Heart weights may be obtained for a number of rats used in a drug-testing experiment. The Kolmogorov test could be used to see whether weights are consistent with a normal distribution of mean 11 g and standard deviation 3 g. (These may be established values for a large batch of untreated rats from the same source.) The approach would be appropriate if it were uncertain how the drug might affect heart weight.

Forestry

The volume of usable timber per tree may be obtained for 50 randomly selected trees from a mature forest. We may wish to know if it is reasonable to assume volumes are distributed normally. As no mean and variance are pre-specified, the Lilliefors test or one of those mentioned in Section 3.1.3 would be appropriate.

EXERCISES

3.1 Are the insurance data given in Exercise 2.1 likely to have come from a normal distribution? Test using the Lilliefors test.

3.2 Test whether the data in Exercise 2.1 may have come from a uniform distribution over the interval (300, 6000).

3.3 Assuming symmetry, carry out a test relevant to the situation and data in Exercise 2.5. using a normal scores procedure.

3.4 Use the Lilliefors test to determine whether the second set of sample values in Example 3.5, i.e. 7, 11, 5, 14, 10, 8, 12, 113, 19 is consistent with a normal distribution.

3.5 Using the data in Example 3.6, carry out a test of the hypothesis $H_0: \mu = 94$ against $H_1: \mu \neq 94$ using a randomization test based on deviations from 94.

3.6 Using the data in Example 3.6 carry out a test of the hypothesis $H_0: \mu = 92$ against $H_1: \mu \neq 92$ using a randomization test based on deviations from 92.

3.7 The negative exponential distribution with mean 20 has the cumulative distribution function $F(x) = 1 - e^{-x/20}$, $0 \leqslant x \leqslant \infty$. Use a Kolmogorov test to test the hypothesis that the sample of excess parking times in Exercise 2.12 may come from a population having this distribution.

3.8 In Section A7.2 one set of ages is for 21 members of the McGamma clan. Perform an appropriate test to determine whether it is reasonable to suppose clan McGamma ages at death are normally distributed.

3.9 Use (i) the Wilcoxon signed rank test and (ii) van der Waerden normal scores, to test the hypothesis that the median length of 12-year-old turbot is 73.5, using the data in Exercise 2.13. (*Note*: appropriate van der Waerden scores for tied ranks are determined by using tied rank values of r in the expression $\frac{1}{2}[1 + r/(n + 1)]$ for the relevant quantile.)

4
Methods for paired samples

4.1 COMPARISONS IN PAIRS

Single samples are useful for illustrating basic ideas, but cover only limited applications. More usually, we are faced with (i) two or more independent samples, or (ii) two or more related samples. In this and the next chapter we limit discussion to two samples, dealing in this one with paired samples, a special class of related samples for which many problems reduce to a single-sample equivalent.

4.1.1 Why pair?

Responses of individuals to stimuli often vary greatly, so to compare two stimuli there is clearly a case for applying them both, if practicable, to each of, say, n individuals, or at least to n pairs of individuals as like as possible, and to look at the response differences within individuals or between members of each pair.

Care is needed when subjecting the same individual to two stimuli, since one stimulus may affect the individual's response to the other. For example, if one wants to test two drugs for the alleviation of pain, if the second is given before the effects of the first are completely worn off one could not be sure if any improvement is the same as would have been obtained had the second been given quite independently, or whether the drugs may be affected in their action by the presence of each other ('interaction' is the technical statistical term). Previous administration of one drug might make the other more (or less) effective than it would be on its own; in an extreme, but not uncommon, case administration of one drug after the other might make the patient's condition worse, whereas if either is given separately the effect may be beneficial. Warnings against drug/alcohol interactions are commonly given by doctors.

If it is known that the analgesic effect of a drug lasts only a few hours and the drug is completely eliminated from the body within 24 hours, it might be reasonable to test each drug on the same patient at intervals, say, three days apart and see which gives the greater pain relief. A measure of relief must be recorded. It might be a physical measurement, but for pain relief is more likely to be simply patient preference, e.g. first drug gave greatest relief, second drug gave greatest relief, or both equally good (or equally ineffective). One may score 'drug A gave more relief' as a plus, 'drug B gave more relief' as minus and 'no difference' as zero; then the individual patient scores – plus, minus or zero – provide data for a sign test. If, among n patients, t assess the drugs as equally effective (score zero), then there will be $n - t$ responses that are either plus or minus. If neither drug is consistently superior, the pluses and minuses will have a binomial distribution with parameters $n - t$ and $p = \frac{1}{2}$. We may test the null hypothesis H_0: no preference, against H_1: one drug or the other more commonly preferred.

To avoid either patient or physician prejudices influencing responses such clinical trials are usually carried out using the **double blind** procedure. Neither the patient nor the nurse or doctor administering the drug knows the order of presentation. Drugs are identified by codes available only to research workers who have no direct contact with the patient.

4.1.2 Some further examples

Geffen, Bradshaw and Nettleton (1973) wanted to know whether certain numbers presented in random order to individuals were perceived more rapidly in the right or left visual field, or whether there was no consistent difference, it being a matter of chance whether an individual responded more rapidly in one field or the other. Details of the experiment are given in the original paper and it was also discussed by Sprent (1977). For each of 12 subjects the mean response times to digital information in the left (LVF) and right visual fields (RVF) were measured. Response times in either field varied markedly between individuals; this variation was much greater than that between response times in the LVF or RVF for any one individual.

The data and the differences, LVF − RVF for each individual are given in Table 4.1.

Table 4.1 Mean response time (ms) to digital information

Subject	1	2	3	4	5	6	7	8	9	10	11	12
LVF	564	521	495	564	560	481	545	478	580	484	539	467
RVF	557	505	465	562	544	448	531	458	560	485	520	445
LVF − RVF	7	16	30	2	16	33	14	20	20	−1	19	22

It is clear from Table 4.1 that the response time for all but one individual is faster in the right visual field. The differences are all less than 33 ms – compared with differences of 100 ms or more between individuals within either visual field. Without matching in pairs, these latter differences might well swamp the relatively consistent differences for individuals. We explore this in Exercise 5.8.

Educationists often compare two examination or teaching methods by pairing students so that each member of a pair is as similar as possible in age, intelligence, and previous knowledge of the subject. Each member of a pair is then exposed to one technique and the results of tests interpreted in terms of pairwise differences.

Consistency between two examiners may be compared by asking each to mark the same series of essays without disclosing the marks each awards to one another. The differences between the mark awarded by each examiner for each essay are then compared to see whether one examiner consistently awards a higher mark or whether there appears to be a purely random difference between marks.

An animal husbandry consultant may compare two diets using sets of twin lambs; one diet is fed to each twin, and their growth measured over a period: attention is focused on growth differences between twins in each pair. Twins are used because of genetical similarity. Each of a pair of twin lambs fed on identical diets would tend to grow at much the same rate: when fed different diets any reasonably consistent differences in growth may be attributed to the effect of diet.

In all the above examples the aim of pairing is to make conditions, other than the treatment or factor under investigation, as similar as possible within each pair; the mean or median differences provide a measure of treatment effect, being indicative of a 'shift' in the distribution.

4.1.3 Single-sample analyses of matched pairs

We introduce no new tests in this chapter. Differences between paired observations provide a single sample that can be analysed by the methods developed in Chapters 2 and 3. We must, however, consider the assumptions about observations on each member of the pair and any further precautions in experimental procedure needed to validate the tests. These points are best brought out by examples.

Example 4.1

The problem. Using the LVF, RVF data in Table 4.1 determine if there is evidence of a consistent response difference between the two fields for individuals. Obtain and compare appropriate 95% confidence intervals for that difference.

Formulation and assumptions. We denote the observations on subject i by x_i in the RVF and by y_i in the LVF. We may write our observations (x_1, y_1), $(x_2, y_2), \ldots, (x_n, y_n)$. We analyse the differences $d_i = y_i - x_i$. The d_i are a sample of n independent observations since each is made on a different individual. Under the null hypothesis of no location difference between x_i and y_i the d_i are equally likely to be positive or negative and a sign test is justified. If we can assume symmetry of the d_i under H_0 we may use the Wilcoxon signed rank (or if preferred a normal scores test). There are several different patterns of response distributions that will give a symmetric distribution of positive and negative differences under H_0: the median of the differences is zero. In particular, if we assume response times are identically and independently (but not necessarily symmetrically) distributed in each of the two fields for any individual (but this distribution need not be the same for every individual) then the difference for each individual will be symmetric about zero (since if X and Y have the same distribution, $X - Y$ and $Y - X$ will each have the same distribution and therefore must be symmetrically distributed about zero. We may also get a symmetric distribution of differences if the LVF and RVF for an individual have different symmetric distributions for any individual, providing each of these symmetric distributions has the same mean. It is well to be aware of such subtleties, but by far the most usual situation where a Wilcoxon test is justified is when we may assume identical distributions for LVF and RVF under the null hypothesis; the alternative hypothesis of interest is usually that there is a shift in location only, as indicated by a shift in the median of those otherwise identical distributions in one field or the other. Often the d_i values will themselves indicate if there is serious asymmetry.

If we are willing to assume also that the differences are approximately normally distributed, a t-test is appropriate. We may test whether normality is a reasonable assumption by the Lilliefors test or one of its alternatives.

We consider relevant methods in turn.

Procedure. The procedures applied to differences are essentially those for one sample, so we only sketch details. The reader is urged to carry through the full calculations as an exercise.

The differences in Table 4.1 may be ordered and are:

$$-1, 2, 7, 14, 16, 16, 19, 20, 20, 22, 30, 33$$

and the corresponding signed ranks are

$$-1, 2, 3, 4, 5.5, 5.5, 7, 8.5, 8.5, 10, 12, 12$$

For the sign test for zero median, we have 1 minus and 11 pluses. From Table A1 this is clearly highly significant in a two-tail test at nominal 1% levels (actual level 0.6%). The 95% confidence interval based on the sign test is $(7, 22)$ since we reject H_0 at the 5% level only for 2 or fewer or 10 or more minus signs. For the Wilcoxon test $S_n = 1$ and this is clearly significant also at the 1% level since S_n is well below the critical level of 7 given in Table A2. With such a clear-cut result we need not make special allowance for the two pairs of ties in the ranks.

A 95% confidence interval based on the Wilcoxon test is $(9.5, 23.5)$. We give the full tableau for determining this in Table 4.2 although, as explained in Section 2.2.6, only a few terms in the upper left and lower right need be calculated if this is done manually. If a computer program is available it is worth getting the whole tableau, as the Hodges–Lehmann estimator of the median is then easily read (or computed). There are 78 entries in the body of Table 4.2, so the median is mid-way between the 39th and 40th ordered entry, and has the value 17.25. If only a pocket calculator is used we do not need the complete tableau even if we want a point estimator (see comment 4 on Example 2.10).

Table 4.2 Paired means for Wilcoxon confidence limits

	− 1	2	7	14	16	16	19	20	20	22	30	33
− 1	− 1	0.5	3	6.5	7.5	7.5	9	9.5	9.5	10	14.5	16
2		2	4.5	8	9	9	10.5	11	11	12	16	17.5
7			7	10.5	11.5	11.5	13	13.5	13.5	14.5	18.5	20
14				14	15	15	16.5	17	17	18	22	23.5
16					16	16	17.5	18	18	19	23	24.5
16						16	17.5	18	18	19	23	24.5
19							19	19.5	19.5	20.5	24.5	26
20								20	20	21	25	26.5
20									20	21	25	26.5
22										22	26	27.5
30											30	31.5
33												33

If we assume normality, the t-test value is again significant at the 1% (and indeed at the 0.1% level) in a two-tail test. The 95% confidence interval is (10.05, 22.95) and the point estimate of the mean is the sample mean 16.5. The normal theory test does best and the fact that confidence intervals are not seriously displaced for the sign test indicates a reasonable degree of symmetry. A Lilliefors test may be used to quell any doubts about approximate normality. The test statistic is 0.151. Table A4 indicates this is well below the value required for significance.

Conclusion. We reject the null hypothesis and conclude that response times in the RVF are significantly faster. Consistency of confidence intervals using three different approaches suggests the mean improvement is between about 10 and 22 ms. The parametric normal theory test is justified for these data in the light of the result of the Lilliefors test.

Comments. 1. Note that the set of differences we compare are independent (a necessary condition for validity of our location tests) as each difference is calculated for a different individual.

2. If response rate in the LVF had been measured before that in the RVF for all subjects, a difficulty in interpretation would have arisen. The result might simply imply that there was a learning process and people responded more quickly in the RVF because they were learning to react more quickly – or there could be a mixture of a learning effect and a more rapid response in the RVF. There are two ways of avoiding this difficulty. One is to decide at random which field – left or right – is tested first; that was done in this experiment. This should balance out and annul to a large degree any learning factor. The other more systematic approach is complete balance – select six subjects (preferably randomly) to be tested first in the left visual field; the remaining six are then tested first in the right visual field. Balanced designs provide a basis for separation of a learning effect from a basic difference between field responses, although a rather larger experiment would be needed to determine this by appropriate parametric or nonparametric methods.

A slight modification only is needed to reduce the test that the mean location difference is some specified value d, to one that the mean difference is zero for either the matched pair sign test or Wilcoxon signed rank test.

Example 4.2

The problem. Eleven children are given an arithmetic test; after three weeks' special tuition they are given a further test of equal difficulty (something we say more about in the comments below). Their marks in each test and the individual differences are given in Table 4.3. Do these indicate that the average improvement due to extra tuition is 10 marks?

Table 4.3 Marks in two arithmetic tests

Pupil	A	B	C	D	E	F	G	H	I	J	K
First test	45	61	33	29	21	47	53	32	37	25	81
Second test	53	67	47	34	31	49	62	51	48	29	86
Second − first	8	6	14	5	10	2	9	19	11	4	5

Formulation and assumptions. The question essentially asks whether, if these children can be regarded as a random sample from some hypothetical population (perhaps children of the same age trained in the same educational system or studying the same syllabus), is it reasonable to suppose the median mark difference is 10?

Procedure. If we subtract 10 from the differences given in the last line of the table our problem reduces to that of testing whether the reduced differences are consistent with a population median of zero, providing we assume a shift in the median is the only difference in the distributions for each individual between the two tests.

Our new set of differences is

$$-2, -4, 4, -5, 0, -8, -1, 9, 1, -6, -5$$

If we use a Wilcoxon test we have a zero and ties in profusion so it is safest to allow for these and use the modified large-sample test explained in Section 2.2.4. The differences, arranged in order of magnitude, with appropriate signs are

$$0, -1, 1, -2, -4, 4, -5, -5, -6, -8, 9$$

and the ranks without signs are:

$$1, 2.5, 2.5, 4, 5.5, 5.5, 7.5, 7.5, 9, 10, 11$$

Using the rule given in Section 2.2.4 for rank signs with zeros and ties, the signed ranks become

$$0, -2.5, 2.5, -4, -5.5, 5.5, -7.5, -7.5, -9, -10, 11$$

Here $d_0 = 1$, $d_1 = d_2 = d_3 = 2$ and $S_p < S_n$ and $S_p = 19$. Using the modified form of (2.4) we find

$$Z = (19 - 32.5)/(11.70) = -1.15$$

This is well below the value required for significance at the 5% level even in a one-tail test (which would hardly be appropriate without further evidence that any improvement must be at least ten).

Conclusion. We do not reject the hypothesis that the median improvement in score might be 10.

Comments. 1. Changing our hypothesis from H_0: median difference is 10, to H_0': median difference is zero is equivalent to the single-sample test that the median for differences is 10, in which we calculate our test statistic as deviations from 10. Note that the differences $-2, -4, 4, -5, \ldots$ are simply the deviations from 10 of the pupil test score differences given in Table 4.3.

2. A conscientious statistician may query how one decides whether two arithmetic tests are of equal difficulty. Might not improved marks in the second test imply that it was in some sense easier? The most the statistician can do here is seek the assurance of the educationist conducting the tests that reasonable precautions have been taken to ensure equal difficulty. Sometimes standard tests that have been tried on large groups of students with results that

show pretty convincingly they are of equal difficulty are used in such situations.

3. Is it realistic to test simply for a shift in median? Sometimes pupils who are good to start with may benefit little from extra tuition; very bad pupils might likewise show little benefit from extra tuition. It may only be those in the middle of the ability range that show appreciable benefit. The statistician may find evidence of this by looking at data – and there are indeed tests for such tendencies. If such effects are present we would be wise to check with the experimenter just what is the aim of extra tuition. For the data in this example there appears to be no such evidence, but deciding what should be tested or estimated is often a matter for fruitful discussion between statistician and experimenter.

4.2 A LESS OBVIOUS USE OF THE SIGN TEST

A test may sometimes be appropriate although the way the data are presented does not immediately suggest its relevance.

Example 4.3

The problem. Members of a mountaineering club argue interminably as to which of two rock climbs is the more difficult. Seeking evidence to settle the argument one member checks the club log book. This records all climbs by members; simply stating whether or not they were successfully completed. He checked records for all 108 members who had attempted both climbs and noted whether they had succeeded or failed in each case. His findings are summarized in Table 4.4.

Table 4.4 Outcomes of two rock climbs

| | | First climb | |
		Succeeded	Failed
Second climb	Succeeded	73	14
	Failed	9	12

Is there reasonable evidence that one climb is the more difficult?

Formulation and assumptions. What evidence is relevant to any difference in severity of the two climbs? A moment's reflection shows that if a climber has succeeded at both climbs this in itself gives no information about which is the harder. If a climber fails on both climbs this also gives no indication of relative difficulty. Success on both or failure on both effectively are **ties** so far as comparing difficulty is concerned. If we had additional information about each climber's assessment of the difficulty or the time taken for each, the situation would be different. As it is, our meaningful comparators of difficulty are numbers who succeed on one but fail on the other climb. From the above table we see that 9 succeeded on the first climb but failed the second. We may think of this as a plus for the first climb; also 14 failed on the first climb but succeeded on the second. We may think of this as a minus for the first climb.

We have here an analogy to the sign test situation. If the first climb is the easier we expect more to succeed on it but fail on the second. If the second were the easier we would expect more to fail on the first but succeed on the second, among those who succeed on one and fail on the other. Those who succeed or fail on both are irrelevant to assessing relative difficulty. Regarding success on first and failure on second as a plus, and the opposite result as a minus, if the climbs are of equal difficulty (which we take as the null hypothesis) the number of pluses has a binomial distribution with $p = \frac{1}{2}$, and where n is the total number of pluses and minuses.

Procedure. There are 9 pluses and 14 minuses, so $n = 23$. Using the normal approximation to the binomial we find, using equation (2.2), that $z = (9.5 - 11.5)/\sqrt{(23/4)} = -0.83$, or $|z| = 0.83$, well below the magnitude required for significance.

Conclusion. We cannot reject the null hypothesis that the climbs are of equal difficulty.

Comment. The fact that $73 + 12 = 85$ out of a total of 108 observed pairs provides no information on the relative difficulty of the climbs may seem wasted data, but the proportion of 'wasted' data does give some indication of how big or small the difference is. There are many situations where we require a large number of observations simply because we are looking for a small difference which may not be clearly distinguishable when the only relevant criteria are success–failure or failure–success categories.

The test just described is one form of a test known as McNemar's test. We say more about this test in Sections 8.1.2 and 8.2.5. Conover (1980, Section 3.5) presents the test in a more formal way that nevertheless reduces in application to what in essence is a paired sample sign test.

4.3 FIELDS OF APPLICATION

In most of the applications listed here, if numerical values of differences in the matched pairs are available and do not appear to be too skew, the Wilcoxon test (or an analogous test using normal scores) is likely to be appropriate. The matched pair t-test becomes appropriate when there is approximate normality in the differences $d_i = y_i - x_i$. This may sometimes be the situation even when the distributions of X, Y are far from normal.

Laboratory instrument calibration

Two different kinds of instrument reputedly measure the same thing (e.g. hormone level, blood pressure, sugar content of urine, bacterial content of sputum), but each is subject to some error. Samples from, say, each of 15 patients, might be divided into two sub-samples, the first being analysed in one instrument, the second in the other. A Wilcoxon procedure is appropriate to detect any systematic difference between instruments. When measuring the same thing, a systematic difference in the means or medians is often referred to as **mean** or **median bias.** The term 'bias' alone is usually taken to imply mean bias; mean and median bias may not coincide if there are other distributional differences.

Biology

Heartbeat rates of rabbits might be observed before and after they are fed a hormone-rich diet. The Wilcoxon test is appropriate to investigate median shift. 'Before' and 'after' measurements are common in many medical and biological situations, including experiments on drugs and other stimuli, which may be either physical or biological (e.g. a rabbit's blood pressure may be measured while it is on its own and again after it has shared a cage for half an hour with a rabbit of the opposite sex).

Medicine

An instrument called a Vitalograph is used to measure lung capacity. Readings may be taken on workers at the beginning and end of a shift to study any effect on lung capacity of fumes inhaled in some industrial process; or on athletes before and after competing in a 100 metre sprint.

Agriculture

In a pest control experiment each of 10 plots may contain 40 lettuce plants. Each plot is divided into two halves; one half chosen at random is sprayed with one insecticide; the second with another. Differences in number of uninfested plants can be used in a Wilcoxon test to compare the effects of the insecticides. Incidentally, pest control experiments are a situation where a normality assumption is often suspect.

Trade

Two different check-out systems are being compared in a supermarket. At randomly selected half-hour periods during a week the numbers passing through each system are recorded. The data might provide information for a

Wilcoxon test to detect any differences in processing speeds. This would probably only be appropriate if the queuing systems were busy throughout each test period. If there were slack periods this might influence customer choice as to which type of check-out point they went to. If neither queue were busy throughout the period it might be better to base the analysis on mean processing time per person for each period.

Psychology

For sets of identical twins, if it were known which were the first-born, times to carry out a manual task may be compared for each pair to see if there is any indication that the first-born is in general quicker. The choice may well be between a Wilcoxon test and a t-test.

Road safety

Drivers' reaction times in dangerous situations may be compared before and after each has consumed a given amount of alcohol.

Space research

Potential astronauts might have the enzyme content of their saliva determined before and after they have been subjected to a zero gravitational field in a simulator. Such biochemical evidence is important in determining physiological reactions to space travel.

Education

To decide which of two examination questions is the harder each may be given to 150 candidates and a record taken of numbers who complete both, complete neither, complete the first but not the second or the second but not the first. Numbers in the latter two categories can be used in a McNemar sign test for evidence of different levels of difficulty.

EXERCISES

4.1   Blood pressures of 11 patients are measured before and after administration of a drug known not to raise blood pressure, but which might have the effect of lowering it. The differences in systolic blood pressure (pressure before – pressure after) for each patient are:

$$7 \quad 5 \quad 12 \quad -3 \quad -5 \quad 2 \quad 14 \quad 18 \quad 19 \quad 21 \quad -1$$

Use an appropriate nonparametric test to see if the sample (assumed effectively random) contradicts the hypothesis H_0: no systematic change in blood pressure.

4.2   Samples of cream from each of ten dairies (A to J) are each divided into two portions. One portion from each is sent to laboratory I and the other to laboratory II with the request that they do bacterial counts. The counts (thousands of bacteria per ml) are as follows:

Dairy	A	B	C	D	E	F	G	H	I	J
Lab I	11.7	12.1	13.3	15.1	15.9	15.3	11.9	16.2	15.1	13.6
Lab II	10.9	11.9	13.4	15.4	14.8	14.8	12.3	15.0	14.2	13.1

Use the Wilcoxon signed rank method to test whether the mean difference between laboratories for sub-samples from the same dairies differ significantly. Obtain also 95% and 99% confidence intervals for mean or median difference and compare these with the corresponding intervals using the optimal method when normality of differences is assumed.

4.3   A hormone is added to one of otherwise identical diets given to each of 40 pairs of twin lambs. Growth differences over a 3-week period are recorded for each pair and signed ranks were allocated to the 40 differences. The lower rank sum was $S_n = 242$. There was only one rank tie and it is not known *a priori* whether the hormone will increase or decrease the growth rate. Investigate the evidence that the hormone may be affecting growth rate.

4.4   A psychologist interviews both the father and mother of each of 17 unrelated but mentally handicapped children, asking each one individually a series of questions aimed to test how well they understand the problems their child is likely to face in adult life. For each family it is recorded whether it is the father (F) or mother (M) who shows the better appreciation of these potential problems. For the 17 families the findings are

F   M   M   F   F   F   F   F   F   F   M   F   F   F   M   F   F

Has the psychologist sufficient evidence to conclude that by and large fathers have a better understanding of the situation?

4.5   To test people's ability to induce events by paranormal influence directed at specific scenes, a parapsychologist asks nine subjects who each claim psychic powers to use these to induce carelessness in customers visiting a certain china shop so as to increase the breakage rate of items due to accidents as they move round the shop examining the stock. Each subject is asked to 'will' customers to break items on a specified (and different) morning (the test day) and the number of breakages per 500 customers entering the shop on that morning is compared with the number of breakages per 500 customers entering the shop on the corresponding day of the previous week (the control day) when no such attempt to induce breakages had been made. The results for the nine subjects are as follows:

Numbers of items broken per 500 customers entering shop

Subject	1	2	3	4	5	6	7	8	9
Test day	2.5	3.5	1.0	4.6	3.8	5.9	6.1	15.1	2.4
Control day	0	3.7	2.4	3.8	4.1	1.2	1.1	1.3	0

Analyse these results by what you consider the most appropriate parametric or nonparametric methods to show whether or not they indicate that a paranormal phenomenon is exhibited.

4.6 One hundred travelling salesman are asked whether or not they think it is in the interests of public safety to prohibit people drinking any alcohol less than 3 hours before driving a car. They are then given a lecture and shown a video on the relevance of alcohol as a cause of some serious accidents. They are then again asked their opinion on the desirability of such a ban. Do the results given below indicate a significant change in attitude as a result of the lecture and video?

		Before lecture	
		In favour	Against
After lecture	In favour	31	16
	Against	8	45

4.7 A canned-soup manufacturer is experimenting with a change in the formula for his tomato soup. There is a regular tasting panel of 70 who each taste samples of the current product and the new formula product (without being told which is which). Of the 70, 32 prefer the new formula product, 25 the current product and the rest cannot distinguish between the two. Is there evidence to reject the null hypothesis that consumer preference is likely to be equally divided?

4.8 Do the data in Exercise 4.7 support a claim that as many as 75% of those who have a preference may prefer the new soup?

4.9 To produce high-quality steel one of two hardening agents A, B, may be added to the molten metal. As the hardness of steel varies from batch to batch, to test the two agents batches are sub-divided into two portions and one hardening agent is added to one portion and the other to the second portion to produce pairs. To compare hardness, sharpened specimens for each pair are used to make scratches on each other: that making the deeper scratch on the other is regarded as the harder specimen. For 40 batch pairs, B is adjudged the harder in 24 cases and A in 16. Is this sufficient evidence to reject a null hypothesis of equal hardness?

4.10 For a sub-sample of 10 pairs of steel featured in Exercise 4.9 a more expensive test that produces a hardness index is carried out. The higher the value of this index, the harder the steel. The results were as follows:

Batch no.	1	2	3	4	5	6	7	8	9	10
Additive A	22	26	29	22	31	34	31	20	33	34
Additive B	27	25	31	27	29	41	32	27	32	34

Use an appropriate test or tests to determine whether these data support the conclusions reached in Exercise 4.9.

4.11 On the day of the third round of the Open Golf Championship in 1987 before play started a TV commentator said that conditions were such that the average scores of players were likely to be at least three higher than those for the second round. For a random sample of the 77 players participating in both rounds the scores were:

Round 2   73   73   74   66   71   73   68   72   73   72
Round 3   72   79   79   77   83   78   70   78   78   77

Do these data support the commentator's claim? Consider carefully whether a one- or two-tail test is appropriate.

4.12 Pearson and Sprent (1968) gave data for hearing loss (in decibels below prescribed norms) at various frequencies. The data below show these losses for 10 individuals aged between 46 and 54 at frequencies of 0.125 and 0.25 kc/s. A negative loss indicates hearing above the norm. Is there an indication of a different loss at the two frequencies?

Subject	A	B	C	D	E	F	G	H	I	J
0.125 kc/s	2.5	− 7.5	11.25	7.5	10.0	5.0	7.5	2.5	5.0	8.75
0.25 kc/s	2.5	− 5.0	6.35	6.25	7.5	3.75	1.25	0.0	2.5	5.0

4.13 In any matched pair test in which symmetry is assumed for differences a normal scores test may be used. Apply one to the data in Example 4.2.

4.14 Scott, Smith and Jones (1977) give a table of estimates of the percentage of UK electors predicted to vote Conservative by two opinion-polling organizations in each month in the years 1965–70. For a random sample of 15 months during that period the percentages were:

Organization A   43.5   51.2   46.8   55.5   45.5   42.0   36.0   49.8
Organization B   45.5   44.5   45.0   54.5   49.5   43.5   41.0   53.0

Organization A   42.5   50.8   36.6   47.6   41.9   48.4   53.5
Organization B   48.0   52.5   41.0   47.5   42.5   45.0   52.5

Do these results indicate a significant tendency for one of the organizations to return higher percentages than the other? Obtain an appropriate 95% confidence interval for any location difference between poll
predictions of the Conservative vote during the period in question.

5

Tests and estimation for two independent samples

5.1 LOCATION TESTS AND ESTIMATES

When we have two independent samples and wish to compare their location (medians or means) we can no longer reduce this to a one-sample problem.

5.1.1 The median test

For two independent samples the sign test extends to the median test. If both samples are from populations with the same median then the sample medians are each appropriate estimators of the common population median. If we combine the two samples and determine the median of this combined sample it again is an appropriate estimate of the population median under H_0: the populations have the same median. If we denote this joint sample median by M then the number of members of the first and second sample with values above M will both have a binomial distribution with $p = \frac{1}{2}$. If, on the other hand, the samples come from populations with different medians, we expect one sample to have a preponderance of values above M, the other to have a preponderance of values below M. If $a_i, i = 1, 2$ are the number of observations exceeding M in sample i of size n_i then, excluding the case where any observation equals M we may set out the numbers above and below M for each sample in a 2×2 table:

	Sample 1	Sample 2
Above M	a_1	a_2
Below M	$n_1 - a_1$	$n_2 - a_2$

This table of counts categorized by the characteristics **sample number** and **position relative to** M is an example of a contingency table of counts and is a special case of a situation we consider in Section 8.2.4 where we develop the appropriate test in more detail. This is a good example of how nonparametric tests may sometimes be applied in several situations that at first sight are not obviously related to one another. At this stage we merely record that one formulation of the test statistic is

$$T = (2a_1 - n_1)^2(n_1 + n_2)/n_1 n_2 \tag{5.1}$$

which under H_0 has a chi-squared distribution (see Section A5) with 1 degree of freedom providing n_1, n_2 are not very small. Values of T in excess of 3.84 indicate significance (rejection of H_0) in a two-tail test at the 5% level.

The test is easy to perform but, like the sign test, is often less efficient than a test using more of the information in the data. We give an example of its application in comment 2 on Example 5.2. Note that the null hypothesis only requires each population to have the same median; they may differ in other characteristics.

Tukey's quick test described by Sprent (1981, pp. 129–32) is even simpler than the median test but is not recommended as it has low power against most alternatives likely to be of interest.

5.1.2 The Wilcoxon–Mann–Whitney test

Methods based on ranks and associated randomization tests are widely used. The best known of these are the essentially identical but differently expressed Wilcoxon rank sum test (related to, but distinct from, the signed rank test) and the Mann–Whitney test, developed independently by Wilcoxon (1945) and Mann and Whitney (1947). We follow the modern trend of referring to both versions as the Wilcoxon–Mann–Whitney test. The Wilcoxon format reflects the basic theory more directly but many people find the Mann–Whitney approach easier to apply. Each leads to equivalent tests.

The general philosophy has much in common with the Wilcoxon signed rank test; instead of distinguishing ranks by sign and then summing those of like sign, we make a joint ranking of observations from the two samples and sum the ranks associated with one sample. If both samples come from the same population we expect a fair mix of low-, medium- or high-ranking observations in each sample. If the alternative to a null hypothesis of identical populations is that our samples come from populations with distributions differing only in location (i.e. mean or median), then we might reasonably expect lower ranks to dominate in one population and higher ranks in the other. This concept of a shift in location is important as it epitomizes a measure of an additive treatment effect (or a 'constant' treatment difference). To establish a critical region when we base our test on ranks we first calculate the probability of each rank sum for all possible associations of ranks with samples of given size and reject the null hypothesis if our result falls in a set that has low probability under the null hypothesis, but is more likely under the alternative.

Example 5.1

Consider a simple case in which we have nine observations ranked 1 to 9. To relate this to two samples of 4 and 5 observations we take a random sample of 4 of these (the first sample); the 5 values not selected then correspond to the second sample ranks. Physically the situation is analogous to drawing 4 tickets from a hat containing 9 tickets labelled 1 to 9. Draws like (1, 5, 6, 8) or (2, 3, 7, 9) or (2, 4, 5, 7) would not surprise us.

We may be surprised if we draw (1, 2, 3, 4) or (6, 7, 8, 9). But why? If we are selecting at random any sample of four has an equal probability of selection. Our surprise stems from the fact that the last two samples contain respectively all the low and all the high ranks, whereas the others contain a good mix of ranks. This is reflected in the rank sums. There are many more samples with rank sum 17, say, than there are with rank sums 10 or 30; (1, 2, 3, 4) is the only one with rank sum 10, but (1, 3, 6, 7) or (2, 4, 5, 6) or (1, 2, 5, 9) or (1, 2, 6, 8) or (2, 3, 5, 7) do not exhaust all those that add to 17. Can you find at least one other? This property of the sums reflects the fact that samples with a good mixture of ranks are more common than those with extreme ranks only. If, on the other hand, the samples came from populations differing in location, extreme ranks associated with one sample are clearly more likely.

To associate probabilities with the various rank sums we first calculate how many different samples of 4 we can get from 9. This is a combinations problem (see Section A2). The number of ways we can select 4 items from 9 without replacement and without regard to order is the binomial coefficient.

$$^9C_4 = \frac{9 \times 8 \times 7 \times 6}{1 \times 2 \times 3 \times 4} = 126$$

All 126 selections are equally likely, so each has a probability of $1/126$. Since (1, 2, 3, 4) is the only sample with sum 10, this sum has a probability of $1/126$. A sum of 11 arises only with (1, 2, 3, 5) so it also has a probability of $1/126$. A sum of 12 is obtained with (1, 2, 3, 6) or (1, 2, 4, 5); the associated probability is therefore $2/126$. The probability of a sum 12 or less is therefore $4/126$ (adding the three relevant probabilities).

If we are testing H_0: the samples are from the same population, against H_1: the samples are from populations differing in location only (i.e. in median or mean), then a two-tail test is appropriate. If we specify in H_1 the direction of any possible shift (i.e. which sample is supposed to come from the population with a possibly higher location parameter) a one-tail test is appropriate. For a one-tail test a sum for the sample of 4 which does not exceed 12 has probability $4/126 = 0.032$. For a two-tail test we also need the probabilities associated with the highest sums. It is left as an exercise (Exercise 5.1) to show that these are 30, 29, 28 with associated probabilities $1/126$, $1/126$, $2/126$ respectively. The top and bottom tails are symmetric. Thus, if the sum of 4 ranks associated with the smaller sample were 28 and we reject, in a two-tail test, the hypothesis of identical populations we would in fact be testing at an $8/126 = 0.063$ significance level (i.e. with critical region containing the sums 10, 11, 12, 28, 29, 30); the size of this region exceeds the conventional 0.05 used in testing at the 5% level.

In general, for samples of size m, n tables exist for critical values at the 5% and 1% levels both for the rank sums S_m, S_n (conveniently labelled the Wilcoxon statistics) and for the closely related (Mann–Whitney) statistics U_m and U_n that we define below. The latter tables are more common and slightly more versatile in use. Note that here S_n refers to the sample of n observations whereas in previous chapters we used the same notation for the sum of negative ranks. There should be no confusion if one recalls the context in each case.

We calculate S_m or S_n. It does not matter which, but generally speaking it is easier to calculate either the sum from the sample that tends to involve the smaller ranks, or that for the sample with the smaller number of observations if m, n differ substantially. Tables give for various m, n, critical values of S_m, S_n, or of

$$U_m = S_m - \tfrac{1}{2}m(m+1)$$

and

$$U_n = S_n - \tfrac{1}{2}n(n+1)$$

Since $S_m + S_n$ is the sum of all ranks from 1 to m, i.e. $\tfrac{1}{2}(m+n)(m+n+1)$, it is easily verified that

$$U_m = mn - U_n \tag{5.2}$$

so that only one of U_m, U_n need be calculated *ab initio* if we want them both. Tables A5 ($m = n$) and A6 ($m \neq n$) give, for various m, n, the maximum of the lesser of U_m, U_n indicating significance in a one- or two-tail test as appropriate. Values for more extensive combinations of (m, n) are given by Neave (1981, p. 30) who refers to the test as the Mann–Whitney) test. Some published tables give critical values (or sometimes quantiles) for S_m, S_n (e.g. Table A7 in Conover, 1980, gives various quantiles for S_m, S_n).

The Mann–Whitney formulation in Section 5.1.3 leads directly to U_m or U_n. The minimum value of either U_m, U_n is always zero, whereas the minimum of S_m, S_n depends on m, n.

Example 5.2

The problem. In Example 2.6 we gave numbers of pages for a random sample of 12 statistics books from my shelves. Another of my shelves contains a collection of my favourite works of fiction and non-fiction (excluding any on statistics or major reference works). From these 143 general books, I noted the numbers of pages for a random sample of 16. These are given below, arranged in ascending order, together with the value for the 12 statistics books given in Example 2.6.

Numbers of pages in general books:

$$\begin{array}{cccccccc}
29 & 39 & 60 & 78 & 82 & 112 & 125 & 170 \\
192 & 224 & 263 & 275 & 276 & 286 & 369 & 756
\end{array}$$

Numbers of pages in statistics books:

126 142 156 228 245 246 370 419 433 454 478 503

Is there evidence that these samples come from populations differing in median numbers of pages?

Formulation and assumptions. If we assume a location shift is the sole distributional difference, the Wilcoxon–Mann–Whitney test is appropriate.

Procedure. We give in Table 5.1 all sample values in ascending order and the associated ranks. For ease of identification values and ranks for the second sample (statistics books) are underlined.

Table 5.1 Combined ranks for two independent samples

Value	29	39	60	78	82	112	125	126	142	156
Rank	1	2	3	4	5	6	7	8	9	10
Value	170	192	224	228	245	246	263	275	276	286
Rank	11	12	13	14	15	16	17	18	19	20
Value	369	370	419	433	454	478	503	756		
Rank	21	22	23	24	25	26	27	28		

Here $m = 16$ and $n = 12$.

$$S_m = 1 + 2 + 3 + 4 + 5 + 6 + 7 + 11 + 12 + 13 + 17$$
$$+ 18 + 19 + 20 + 21 + 28 = 187,$$

whence

$$U_m = 187 - \tfrac{1}{2} \times 16 \times 17 = 51$$

and

$$U_n = 12 \times 16 - 51 = 141$$

We may check this last result by calculating S_n and hence U_n directly.

From Table A6, or Neave's table of critical values, we see that using a two-tail test at the 5% level, the critical value of U_m is 53.

Conclusion. Our observed $U_m = 51$ is less than the critical value, so we reject the hypothesis that the medians are equal.

Comments. 1. The assumption that the populations are identical apart from location may be suspect in view of the presence of one very long general book with 756 pages that somewhat offsets the general impression of shorter volumes. This suggests that a difference in spread of the two populations is also possible. We examine this possibility in Examples 5.3 and 5.10.

2. Had we used the median test in Section 5.1.1 it is easily seen from Table 5.1 that the combined median is $M = \frac{1}{2}(228 + 245) = 236.5$. For general books 6 out of 16 are above M and for statistics books 8 out of 12 are above M. From equation (5.1) this gives $T = 2.33$, well below the value 3.84 required for rejection of H_0. This suggests a loss of power because we use less sample information.

3. Although the Wilcoxon–Mann–Whitney text is usually used for location shift it is valid for testing some more general hypotheses. See e.g. Conover (1980, pp. 216–17).

With an assumption of both symmetry and normality the t-test becomes a valid competitor to the Wilcoxon–Mann–Whitney test.

Example 5.3

The problem. Test for equality of means using the data in Example 5.2, making the additional assumption that the samples come from normal populations that differ if at all only in location (i.e. with the same parameter σ^2).

Formulation and assumptions. The two-sample t-test is appropriate.

Procedure. The sample means are $\bar{x}_m = 208.5$, $\bar{x}_n = 316.67$ and the test statistic (see Section A4.2) with 26 degrees of freedom is $t = 1.73$.

Conclusion. Using a two-tail test we do not reject the hypothesis that the population means differ. Were a one-tail test appropriate we would, testing at the 5% level, just conclude that the general books had a lower mean, the critical t-value being 1.71.

Comments. 1. The two-tail Wilcoxon–Mann–Whitney test detects a difference which is undetected in a t-test. The key to the t-test breakdown is the observation 756; it is influential in inflating the estimated variance in the denominator of the t-statistic. However, the formal test for equality of variances, when normality is assumed, using the F-statistic (see Section A5) does not lead to rejection of variance equality, so it is not unreasonable to suspect some non-normality in the data. The Lilliefors test of Section 3.1.3 for normality indicates that the sample of 16 general books is not quite inconsistent with normality (the probability of getting the observed value of the relevant statistic is less than 0.10 but greater than 0.05). However, the power of this test is not particularly high against long-tail alternatives or slight skewness. The Shapiro–Wilks test described by Conover (1980, Section 6.2) is more powerful in these circumstances and rejects the normality hypothesis at the 1% level.

2. We describe nonparametric tests for equality of variance in Section 5.4.

3. The Wilcoxon–Mann–Whitney test does better because it plays down the outlying nature of the observation 756. Whereas this is numerically well above 503 – the highest sample value for statistics books – it would have been given the same rank had it been only 504. This reduction in influence of an outlier (compared with the effect on the t-test) is one of the reasons for robustness of the Wilcoxon–Mann–Whitney procedure.

5.1.3 The Mann–Whitney formulation

Mann and Whitney (1947) proposed a direct way of obtaining U_m or U_n which we describe in an example.

Example 5.4

The problem. Recalculate the test statistic for the data in Example 5.2 using the Mann–Whitney approach.

Formulation and assumptions. Again, we order the combined samples as in Table 5.1 but we do not even need to write down the ranks. We observe the number of observations in one sample exceeding each member of the other sample. The total obtained will be either U_m or U_n.

Procedure. The process can be carried out by visually inspecting the 'Values' line in Table 5.1. The relevant values are reproduced in Table 5.2.

Table 5.2 Combined ordering for two independent samples

29	39	60	78	82	112	125	126	142	156
170	192	224	228	245	246	263	275	276	286
369	370	419	433	454	478	503	756		

For example, if we consider general book values (not underlined) exceeding each statistics book value (underlined) we note that 126 is exceeded by 170, 192, 224, 263, 275, 276, 286, 369, 756 from the other sample – a total of 9 values in all. Obviously 142 and 156 are each exceeded by the same 9 values. Similarly 228, 245 and 246 are each exceeded by 6 values – 263, 275, 276, 286, 369, 756; and finally each of 370, 419, 433, 454, 478, 503 is exceeded by the single value 756 only. It is easily verified that this gives a total of 51 excesses, equal to U_m. The test is now completed as in Example 5.2.

Conclusion. As in Example 5.2.

Comment. We do not prove it, but this equivalence of the methods is general.

Many people feel the Mann–Whitney procedure is easier to apply; it is easy to program for a computer, and we shall see in Section 5.2 that it is relevant to determining a confidence interval. We described the Wilcoxon procedure first because it better exhibits the rationale of the test and shows its relationship to the signed rank test.

5.1.4 A large-sample normal approximation

For values outside the range of convenient tables – generally speaking, if one or both of m, n is greater than 20 – a normal approximation with continuity correction is satisfactory. Under the null hypothesis, if U is the smaller of U_m, U_n and there are no ties,

$$Z = \frac{U + \frac{1}{2} - \frac{1}{2}mn}{\sqrt{[mn(m + n + 1)/12]}} \tag{5.3}$$

has an approximately standard normal distribution. We reject at the 5% level in a two-tail test H_0: no location difference, if $Z < -1.96$; other critical values are the usual normal theory ones for appropriate levels in one- or two-tail tests.

Example 5.5

The problem. Apply the approximation in equation (5.3) to Example 5.4.

Formulation and assumptions. Substitute the appropriate values $U_1 = 51$, $m = 12$, $n = 16$ in (5.3).

Procedure. We easily find

$$Z = (51.5 - \tfrac{1}{2} \times 12 \times 16)/[\sqrt{(12 \times 16 \times 29/12)}] = -2.07$$

Conclusion. Since this is less than -1.96 (i.e. $|Z| > 1.96$), we reject H_0 at the 5% level, as we did with the exact test.

5.1.5 Ties

As with the signed rank test, if there are only a few ties, the mid-rank method and the standard test or the normal approximation in (5.3) are unlikely to be misleading. If there are many ties the large-sample approximation (5.3) should be modified as follows.

The continuity correction of $\frac{1}{2}$ in the numerator is deleted and the expression $mn(m + n + 1)/2$ in the denominator has $mn\sum_i(d_i^3 - d_i)/[12(m + n)(m + n - 1)]$ subtracted from it, where the number, r, of d_i terms equals the number of tie groups occurring and d_i equals the number of observations in the ith tie group, i.e. $d_i = 2$ for a tied pair, 3 for a tied triplet, 4 for a tied quadruplet, etc., as in Section 2.2.4.

Example 5.6

The problem. We consider data for age at death for members of two more clans – the McBeta and McGamma – in the Badenscallie burial ground (see Section A7.2). Arranged in ascending order the ages are:

McBeta:	0	19	22	30	31	37	55	56	66	66	67	67
	68	71	73	75	75	78	79	82	83	83	88	96

McGamma:	13	13	22	26	33	33	59	72	72	72	77	78
	78	80	81	82	85	85	85	86	88			

Use the Wilcoxon–Mann–Whitney large-sample approximation to test the hypothesis that these may be regarded as samples from identical populations against the hypothesis that the populations differ in location.

Formulation and assumptions. As the sample sizes are 24, 21 respectively the large-sample approximation will suffice. Ties are replaced by their mid-rank. If there are only a few ties the adjustment to the denominator given above will have little effect.

Procedure. In Table 5.3 we give the joint ranking (with appropriate mid-ranks for ties). Ages and ranks for clan McGamma are underlined.

Table 5.3 Joint ranks for age at death samples

Age	0	13	13	19	22	22	26	30	31	33	33	37
Rank	1	2.5	2.5	4	5.5	5.5	7	8	9	10.5	10.5	12

Age	55	56	59	66	66	67	67	68	71	72	72	72
Rank	13	14	15	16.5	16.5	18.5	18.5	20	21	23	23	23

Age	73	75	75	77	78	78	78	79	80	81	82	82
Rank	25	26.5	26.5	28	30	30	30	32	33	34	35.5	35.5

Age	83	83	85	85	85	86	88	88	96			
Rank	37.5	37.5	40	40	40	42	43.5	43.5	45			

If we use the Mann–Whitney counting method to evaluate U_m or U_n directly some care is needed in counting ties, which are scored as $\frac{1}{2}$. Denoting by U_m the total number of McBeta ranks exceeding McGamma ranks, we find $U_m = 216.5$. Using (5.1) it is easily verified that $U_n = 287.5$.

Using the large-sample approximation without continuity correction and with no denominator allowance for ties, we find

$$Z = (216.5 - 252)/\sqrt{1932} = -35.5/43.95 = -0.81$$

Conclusion. Since 0.81 is well short of the critical magnitude 1.96 required for significance in a two-tail test at the 5% level we do not reject the hypothesis of identical population.

Comments. 1. We ignored the denominator correction for ties. Inspection of Table 5.3 indicates paired ties at 13, 22, 33, 66, 67, 75, 82, 83 and 88 and triple ties at 72, 78, 85. Thus there are 9 paired ties and 3 triple ties. Using the correction given just before this example we find $d_i^3 - d_i = 6$ for each paired tie and 24 for each triple tie. Thus $\sum_i (d_i^3 - d_i) = 9 \times 6 + 3 \times 24 = 126$ and the correction term is $(24 \times 21 \times 126)/(12 \times 45 \times 44) = 2.67$.

The revised Z is

$$Z = -35.5/\sqrt{(1932 - 2.67)} = -35.5/43.92 = -0.81$$

Rounded to two significant figures, the value of Z remains the same, so with only a few ties in large samples correction of the variance estimate is not important.

2. Strictly speaking we did not need to assign ranks to carry out the Mann–Whitney procedure; however, we gave them in Table 5.3 as some readers may prefer to add ranks to get S_m and S_n and deduce U_m and U_n. It is tedious to find U_m directly without a computer.

The next example illustrates an application of the Mann–Whitney procedure to a situation where ties dominate. It is an interesting application to what is essentially grouped categorical data (Chapter 8) where there is a definite ordering of categories.

Example 5.7

The problem. Suppose that instead of full data for some general and statistics books we have only a record of the numbers of pages in the broad bands: (i) under 100, (ii) 100–199, (iii) 200–299, (iv) 300–399, (v) 400–499, (vi) 500 or more. For a sample of 24 statistics books (those in Example 2.1) and 16 general books (those in Example 5.2) the numbers in the categories are as follows:

Pages	< 100	100–199	200–299	300–399	400–499	500 +
Statistics books	0	4	5	8	3	4
General books	5	4	5	1	0	1

Can we accept the hypothesis these are from the same population, or might we suppose they come from populations differing in location?

Formulation and assumptions. We carry out a Wilcoxon–Mann–Whitney test regarding all entries in the same page category as ties. Thus there are 5 ties in < 100, 8 in 100–199, 10 in 200–299, etc. To calculate U_m (excesses of general over statistics books) we note that the first statistics entry is 4 in the 100–199 category. For each of the 4 there is a tie with 4 general books (scored as $4 \times \frac{1}{2}$) and there are $(5 + 1 + 0 + 1)$ higher-ranked general books, each scored as 1. Thus the total contribution to U_m from the 100–199 page grouping is

$$4(\tfrac{1}{2} \times 4 + 5 + 1 + 1) = 4 \times 9 = 36$$

Similarly, the contribution from the category 200–299 pages is $5(\tfrac{1}{2} \times 5 + 1 + 1) = 5 \times 4.5 = 22.5$. It is left as an exercise to confirm that the remaining categories contribute 8×1.5, 3×1 and $4 \times \frac{1}{2}$ respectively, whence

$$U_m = 4 \times 9 + 5 \times 4.5 + 8 \times 1.5 + 3 \times 1 + 4 \times 0.5 = 75.5$$

Similarly we find $U_n = 308.5$. As usual,

$$U_m + U_n = mn = 16 \times 24 = 384$$

In this example $d_1 = 5$, $d_2 = 8$, $d_3 = 10$, $d_4 = 9$, $d_5 = 3$ and $d_6 = 5$, whence, $\Sigma_i(d_i^3 - d_i) = 2478$.

Using the modified form of the variance in the denominator of (5.2) given in this section and ignoring the continuity correction, we find

$$Z = -116.5/\sqrt{(1312 - 50.83)} = -3.28$$

Conclusion. This value is almost significant at the 0.1% level and is certainly highly significant. There is strong evidence for rejecting the hypothesis of identical distributions and for accepting a difference in location.

Comments. The ties here do have an appreciable effect on the denominator. Had we ignored them the value of Z would have been lower in magnitude. Indeed, the effect of ties (since the correction term for them is always positive and is subtracted) is to reduce the denominator in Z. Thus, ignoring ties makes the test more conservative as it lessens the prospect of detecting a significant result when one really exists.

5.2 WILCOXON–MANN–WHITNEY CONFIDENCE INTERVALS

Confidence intervals for a location shift are based on paired differences between samples in a direct extension of the Mann–Whitney approach to hypothesis testing. Effectively what we did in that approach was to make a pairwise comparison between observations for one sample and observations for the other, recording only the number of cases where the second sample observation exceeded that for the first sample. The number of excesses gave U_m or U_n as appropriate. To obtain confidence intervals we require the actual values of extreme differences totalling in number one more than the 'critical' number for rejection of the null hypothesis at the $100\alpha\%$ significance level if we require a $100(1-\alpha)\%$ confidence interval. With a straightforward computer program it is easy to obtain all paired differences. Given these we may get a point as well as an interval estimate. The point estimator, called the Hodges–Lehmann estimator, is the median of the paired differences.

Example 5.8

The problem. For the number of pages data for general and statistics books in Example 5.2 determine a 95% confidence interval for location shift.

Formulation and assumptions. Denoting sample values by $x_1, x_2, \ldots, x_m$ and $y_1, y_2, \ldots, y_n$, we calculate each paired difference $x_i - y_j$. Table A6 gives $U_m = 53$ as the critical value for significance at the 5% level. To determine the confidence interval we reject the 53 smallest and 53 largest differences and the smallest and largest remaining difference give the requisite confidence limits. A point estimate of location difference is the median of all paired differences.

Procedure. It eases computation if the sample values are ordered. Those for one sample may be written across the top row and those for the other down the first column as in Table 5.4. The entries in the body of that table are the differences between the data entry at the top of that column and at the left of that row, e.g. the first entry 97 is $126 - 29$. Note that the difference between each entry in any pair of rows is a constant equal to the difference between the corresponding entries in the left-hand data column, with an analogous constancy for pairs of columns. Note that the largest entries occur in the top right of the table, and that they decrease and eventually change sign as we move towards the bottom left. If using a computer program it may help to have a printout in a format like Table 5.4.

A count shows 51 negative values (see comments); there are two differences of 1; together these constitute the 53 smallest values; the lowest value remaining is 4 and this is the lower 95% confidence limit. A check also shows 53 differences of 241 or more. The largest difference below 241 is 240; this is the upper confidence limit. Since there are in all 192 differences the median lies mid-way between the 96th and 97th ordered difference; it is easy but tedious to verify that these differences are 133 and 134, thus a point estimate of the location difference is 133.5.

Table 5.4 Paired differences for page data

	126	142	156	228	245	246	370	419	433	454	478	503
29	97	113	127	199	216	217	341	390	404	425	449	474
39	87	103	117	189	206	207	331	380	394	415	439	464
60	66	82	96	168	185	186	310	359	373	394	418	443
78	48	64	78	150	167	168	292	341	355	376	400	425
82	44	60	74	146	163	164	288	337	351	372	396	421
112	14	30	44	116	133	134	258	307	321	342	366	391
125	1	17	31	103	120	121	245	294	308	329	353	378
170	− 44	− 28	− 14	58	75	76	200	249	263	284	308	333
192	− 66	− 50	− 36	36	53	54	178	227	241	262	286	311
224	− 98	− 82	− 68	4	21	22	146	195	209	230	254	279
263	− 137	− 121	− 107	− 35	− 18	− 17	107	156	170	191	215	240
275	− 149	− 133	− 119	− 47	− 30	− 29	95	144	158	179	203	228
276	− 150	− 134	− 120	− 48	− 31	− 30	94	143	157	178	202	227
286	− 160	− 144	− 130	− 58	− 41	− 40	84	133	147	168	192	217
369	− 243	− 227	− 213	− 141	− 124	− 123	1	50	64	85	109	134
736	− 630	− 614	− 600	− 528	− 511	− 510	− 386	− 337	− 323	− 302	− 278	− 253

Conclusion. A point estimate of the location difference is 133.5 and a 95% confidence interval for that difference is (4, 240).

Comments. 1. We noted 51 negative differences; this corresponds to the value of U_m calculated by the Mann–Whitney procedure. A moment's reflection shows this is precisely what we must expect, for the method identifies each such difference by sign only, i.e. without assigning a numerical value.

2. If the operation is to be carried out often, especially for large samples, it is well worth having a computer program or package to obtain confidence intervals. The same program automatically allows for hypothesis testing as this requires only a count of the number of positive or negative differences (whichever is the smaller) to obtain the usual test statistic.

3. If we have a complete tableau like that in Table 5.4 other confidence intervals can be obtained. For example, to get a 99% confidence interval, Table A6 shows we need to reject the 41 highest and lowest differences.

4. If no computer program is available a little initiative reduces the calculation required. For instance, to get the lower limit we want the 54th smallest difference. There is no need to evaluate the 51 negative differences; we need only evaluate a few small positive differences and the pattern of the whole tableau in Table 5.4 indicates where these will occur. The general pattern also gives a good indication of approximately where the upper limit will occur.

5. Graphical methods may also be used; these are informative but tedious, so are not popular if a computer is available – indeed even if a pocket

calculator is available. A description of one graphical method is given by Sprent (1981, pp. 144–6).

5.3 TESTS ON FUNCTIONS OF RANKS

As for single samples, randomization tests of location for two independent samples may be based on functions of ranks. One such function is the ordered observations themselves. Here we again face the practical difficulty that the randomization distribution of sums must be calculated afresh for each new data set. There are also difficulties if the samples come from populations differing in dispersion as well as location; a problem that is also well known with the t-test. A more viable possibility is a normal scores approach. Again we confine a detailed account to the van der Waerden scores based on the $r/(m + n + 1)$ standard normal quantiles for $r = 1, 2, \ldots, (m + n)$ in place of ranks themselves. Expected normal scores may also be used and in either case the scores are very like a sample from a normal distribution and we formulate tests accordingly. Symmetry implies that the mean of the $m + n$ van der Waerden scores is zero if there are no ties. If T is the total of the scores associated with either of the samples then, under the null hypothesis, T has mean zero; it can be shown that its variance is approximately $S^2 = nm(\sum_i w_i^2)/\{(m + n - 1)(m + n)\}$ where w_i represents the ith of the $m + n$ van der Waerden scores.

The test statistic, $Z = T/S$, has approximately a standard normal distribution under H_0 and the approximation is good even for relatively small sample sizes in view of the near equivalence of the scores to an 'ideal' sample from a normal distribution.

The approximation that the mean of all w_i is zero is not quite true if there are ties, for in this case we use quantiles based on mid-ranks for the ties, but for only a few ties the effect is negligible as we replace adjacent quantile scores by scores that are close to their mean.

Example 5.9

The problem. Use the data in Example 5.2 to test the hypothesis of identical distributions against that of a location difference using van der Waerden scores.

Formulation and assumptions. The combined samples consist of 28 observations so we replace rank i by the $i/29$ quantile; e.g. if $i = 1$ we require the $1/29 = 0.0349$th quantile. From tables of the standard normal distribution this is easily found to be -1.82. It suffices to record these scores to two decimal places.

Procedure. Table 5.5 gives the ranks in Table 5.1, underlining those for statistics books. Beneath each rank we give the corresponding van der

Waerden score. It is only necessary to obtain the negative values from normal distribution tables; positive scores follow from symmetry.

Table 5.5 Combined ranks and van der Waerden scores for book samples

Rank	1	2	3	4	5	6	7	8	9
Score	-1.82	-1.48	-1.26	-1.09	-0.94	-0.82	-0.70	$\underline{-0.60}$	-0.50
Rank	$\underline{10}$	11	12	13	$\underline{14}$	$\underline{15}$	$\underline{16}$	17	18
Score	$\underline{-0.40}$	-0.31	-0.22	-0.13	$\underline{-0.04}$	0.04	0.13	0.22	0.31
Rank	19	20	21	$\underline{22}$	$\underline{23}$	24	$\underline{25}$	$\underline{26}$	$\underline{27}$ 28
Score	0.40	0.50	0.60	0.70	$\underline{0.82}$	$\underline{0.94}$	$\underline{1.09}$	$\underline{1.26}$	$\underline{1.48}$ $\underline{1.82}$

The sum of the underlined scores is 4.92 (and of those not underlined -4.92). The sum of squares of all scores is 22.515 whence

$$Z = 4.92/\sqrt{[12 \times 16 \times 22.515/(27 \times 28)]} = 2.07$$

Conclusion. Since $Z > 1.96$ we reject the null hypothesis at the 5% significance level.

Comments. $Z = 2.07$ is identical with $|Z|$ in Example 5.5 for the normal approximation to the Wilcoxon–Mann–Whitney test. This is not surprising as the sample sizes are reasonably large. For smaller sample sizes the van der Waerden scores normal approximation may be expected to be better than the normal approximation to the Wilcoxon–Mann–Whitney test. One can obtain exact randomization tests for normal scores, but unless tables are available it is hardly worth working these out for a particular m, n as the normal approximation will suffice for all but very small samples when the Wilcoxon–Mann–Whitney procedure will usually be adequate.

5.4 TESTS FOR EQUALITY OF VARIANCE

Although tests for location shift are more common, we sometimes want to test for heterogeneity of variance; i.e. test H_0: samples are from identical distributions, against H_1: they come from distributions differing in variance. We describe first a simple test that parallels the Wilcoxon–Mann–Whitney location test and uses the same tabulated values for significance.

5.4.1 The Siegel–Tukey test

The test is simple to carry out, but unfortunately not very powerful. The basic idea behind it is that if two samples come from populations differing only in variance, the sample from the population with higher variance will be more spread out with greater extreme values. If we arrange the combined sample in order and allocate the rank 1 to the smallest observation, 2 to the largest, 3 to the next largest, 4 and 5 to the next two smallest, 6 and 7 to the next two largest, and so on, the sum of ranks attached to the population with greater variance should be smaller than if there is no difference in variance. The test will certainly not work well if there is also a location difference. One way to overcome this difficulty if there are indications of a location difference is to 'align' by subtracting from all observations in the sample from the population with higher location an estimate of the location difference (or to add this estimate to observations in the other sample). The variance is unaffected by this location change and the power of the Siegel–Tukey test is increased.

Example 5.10

The problem. Use the Siegel–Tukey test for the samples in Example 5.2 to test for equality of variance.

Formulation and assumptions. We established in Example 5.8 that a point estimate of location difference was 133.5. If we add this to each value for the general book sample we have aligned samples. We apply the Siegel–Tukey test to the aligned samples.

Procedure. After adding 133.5 to the numbers of pages of all general books and arranging the combined samples in ascending order, we obtain the values in Table 5.6. Beneath each value we give rank allocations in the manner described above for the Siegel–Tukey test. We underline values corresponding to statistics books.

Table 5.6 Allocation of ranks for Siegel–Tukey test

Value	126	142	156	162.5	172.5	193.5	211.5	215.5	228	245
Rank	1	4	5	8	9	12	13	16	17	20

Value	245.5	246	258.5	303.5	325.5	357.5	370	396.5	408.5	409.5
Rank	21	24	25	28	27	26	23	22	19	18

Value	419	419.5	433	454	478	502.5	503	889.5
Rank	15	14	11	10	7	6	3	2

For the statistics books (underlined) $m = 12$ and $S_m = 140$, whence $U_m = 62$, above the maximum value for significance at the 5% level even for a one-tail test. Table A6 indicates U_m must not exceed 60 in the latter case.

Conclusion. We do not reject the null hypothesis that the populations have the same variance.

Comments. In Example 5.3 we stated that the normal theory test did not reject the hypothesis of equal variance. Had the observation 756 not occurred in the general book sample we would not have suspected a possible inequality of variance. This might well be regarded as an outlier. In the Siegel–Tukey test this observation is weighted downward, getting no more weight than if the observation had been 370 (which after adding 133.5 for the location adjustment would become 503.5 – the largest joint sample value for aligned samples. In this sense the test is robust. Note that robustness is not synonymous with power. Ideally, we like tests to be both robust and powerful. In practice this may be difficult to achieve here as robust methods tend to play down the influence of extremes and this is often the main discrepancy implying a variance difference.

5.4.2 The squared rank test for variance

If the means of X, Y are respectively μ_x, μ_y then equality of variance implies $E[(X - \mu_x)]^2 = E[(Y - \mu_y)]^2$, where $E(X)$ means the expectation of X.

Conover (1980) proposes a test for equality of variance based on joint ranks of $(x_i - \mu_x)^2, (y_i - \mu_y)^2$. In practice it is unlikely that the population means will be known so it is reasonable to replace them by sample estimates $\bar{x}, \bar{y}$. It is not even necessary to square the deviations to obtain the required rankings, for the same order is achieved by ranking the absolute deviations, i.e. $|x_i - \bar{x}|$ and $|y_i - \bar{y}|$. If we denote the squares of the ranks of these absolute deviations by $u_i(x), u_i(y)$ then the sum, T, of the squared ranks for either sample can be used as a test statistic. If we choose $u_i(x)$ the test statistic is

$$Z = (T - n\bar{u})/S$$

where $\bar{u}$ is the mean of the squares of ranks (i.e. the sum of all squared ranks divided by $m + n$, the total number of observations) and S is an estimated standard deviation calculated from

$$S^2 = \frac{mn\left\{ \sum_i u_i^2 - (m + n)\bar{u}^2 \right\}}{(m + n)(m + n - 1)} \tag{5.4}$$

For reasonably large sample sizes, Z is approximately a standard normal variable, but exact tables for the 'no tie' case giving quantiles of T, and a large-sample approximation are given by Conover (1980, Table A9).

Example 5.11

The problem. Test the data in Example 5.2 for equality of variances using the squared rank test.

Formulation and assumptions. We require deviations of each observation from its sample mean. The absolute deviations are then ranked, the ranks squared and the statistic T, then Z are calculated.

Procedure. Denoting the general book sample values by x_i and those for statistics books by y_j we find $\bar{x} = 208.5$ and $\bar{y} = 316.67$ (Example 5.3). Absolute deviations from the respective means are calculated for each sample, e.g. for the observation 29 in the general book sample the absolute deviation is $|29 - 208.5| = 179.5$.

Table 5.7 gives the ordered deviations for the combined samples and the corresponding ranks and squares of ranks. Values corresponding to statistics books are underlined.

Table 5.7 Deviations, ranks and squares for squared rank test

Deviation	15.5	16.5	38.5	53.3	54.5	66.5	67.5	70.7	71.7	77.5
Rank	1	2	3	4	5	6	7	8	9	10
Square	1	4	9	16	25	36	49	64	81	100
Deviation	83.5	88.7	96.5	102.3	116.3	126.5	130.5	137.3	148.5	
Rank	11	12	13	14	15	16	17	18	19	
Square	121	144	169	196	225	256	289	324	361	
Deviation	160.5	160.7	161.3	169.5	174.7	179.5	186.3	190.7	547.5	
Rank	20	21	22	23	24	25	26	27	28	
Square	400	441	484	529	576	625	676	729	784	

The sum of the underlined squares is $T = 3956$. The mean of the squared ranks for all 28 observations is 275.5. S^2 may be calculated using (5.4) whence $S = 643.7$.

Thus

$$Z = (3956 - 12 \times 275.5)/643.7 = 650/643.7 = 1.009$$

Conclusion. Since $|Z|$ is considerably less than 1.96 we cannot reject the null hypothesis of no difference in variance.

Comments. By using squares of ranks we do not downweigh the outlying observation of 756 so markedly as we did in the Siegel–Tukey test. However, the result is still not significant. Other tests for variance have been proposed by Mood (1954), Moses (1963) and Klotz (1962); details are given by Daniel (1978, Section 3.2) and Marascuilo and McSweeney (1977, Sections 11.11, 11.12).

5.5 A TEST FOR A COMMON DISTRIBUTION

In Section 3.1 we developed tests of hypotheses that a single sample was drawn from a specific distribution. For two independent samples we may want to know if each can reasonably be supposed to come from the same unspecified distribution. The Smirnov test has many similarities to the Kolmogorov test developed in Section 3.1.1.

5.5.1 Smirnov test for a common distribution

The null hypothesis is that the samples come from the same distribution. The alternative is that they come from distributions that have different cumulative distribution functions. We do not specify any particular form for the difference: they might have the same mean but different variances; one may be skew, the other symmetric; they may have the same means and variances, one being symmetric, the other skew; and so on. We compare the sample cumulative distribution functions; the test statistic is the difference of greatest magnitude between these two functions. We illustrate the procedure with an example.

Example 5.12

The problem. Use the Smirnov test to determine if it is reasonable to assume that the two samples in Example 5.2 come from identically distributed populations.

Formulation and assumptions. We compute the sample cumulative distribution functions $S(x)$, $S(y)$ at each sample value and at each of these values we also compute and record the differences $S(x) - S(y)$. For samples of size m, n respectively $S(x)$, $S(y)$ are step functions with steps $1/m$, $1/n$ respectively at each sample value. If there are ties in any one sample there will be a multiple step at the corresponding value.

Procedure. Table 5.8 gives in successive columns the sample values and corresponding values of $S(x)$, $S(y)$ and $S(x) - S(y)$ at each sample point.

The difference of greatest magnitude (final column) is 0.4375 occurring when we introduce the 15th value of 369 in the sample of general books. Table A7 gives critical values for various sample sizes for this test and indicates that with a two-tail test at a nominal 5% level we cannot reject the null hypothesis; with a one-tail test we are just below the critical value at the nominal 5% level. A one-tail test here is essentially a test of the hypothesis that the function for the general book distribution is above that for statistics books at least one x-value, against the alternative that everywhere it is at or below.

Conclusion. We do not reject the hypothesis of identical distributions, but

Table 5.8 Calculation of the Smirnov test statistic

General books x_i	Statistics books y_i	$S(x_i)$	$S(y_i)$	$S(x_i) - S(y_i)$
29		0.0625	0	0.0625
39		0.1250	0	0.1250
60		0.1875	0	0.1875
78		0.2500	0	0.2500
82		0.3125	0	0.3125
112		0.3750	0	0.3750
125		0.4375	0	0.4375
	126	0.4375	0.0833	0.3542
	142	0.4375	0.1667	0.2708
	156	0.4375	0.2500	0.1875
170		0.5000	0.2500	0.2500
192		0.5625	0.2500	0.3125
224		0.6250	0.2500	0.3750
	228	0.6250	0.3333	0.2917
	245	0.6250	0.4167	0.2083
	246	0.6250	0.5000	0.1250
263		0.6875	0.5000	0.1875
275		0.7500	0.5000	0.2500
276		0.8125	0.5000	0.3125
286		0.8750	0.5000	0.3750
369		0.9375	0.5000	0.4375
	370	0.9375	0.5833	0.3542
	419	0.9375	0.6667	0.2708
	433	0.9375	0.7500	0.1875
	454	0.9375	0.8333	0.1042
	478	0.9375	0.9167	0.0208
	503	0.9375	1.0000	0.0625
756		1.0000	1.0000	0.0000

there is a weak indication that the general book distribution function is above that of the statistics book distribution at least one x-value.

Comments. 1. We established in Example 5.2 strong evidence of a location difference, so finding only weak evidence of any difference in the above test may seem disappointing. It is indeed a characteristic of Kolmogorov–Smirnov type tests that they are often less powerful than tests for specific differences such as differences in location.

2. Like the Kolmogorov test, the Smirnov test may appear not to be making full use of the data since it concentrates on the maximum difference. As pointed

out in Section 3.1.1 this is not a serious limitation since the statistic takes in cumulative information. However, if we take all differences into account we may increase the power.

3. The nominal significance levels given in Table A7 may differ from actual significance levels as $|S(x_i) - S(y_i)|$ is a step function, the size of steps depending on sample size. Many published tables for the Smirnov test give not critical values but quantiles which must be exceeded for significance at a given level. Tables A7 and A8 give critical values of $|S(x_i) - S(y_i)|$; Neave (1981, p. 31) gives values of $mn\{\max|S(x_i) - S(y_i)|\}$ for significance for a wide range of sample sizes.

There are difficulties with working out the distribution of statistics for tests that take account of all differences $S(x_i) - S(y_i)$, for clearly these differences are not independent. However, for at least one test that does this, an approximate theory gives simple tests of significance that are virtually independent of sample sizes except for very small samples.

5.5.2 The Cramer–Von Mises test for identical populations

This test is based on the sum of squares of the differences $S(x) - S(y)$ at all sample points. Denoting this sum of squares by S_d^2 the test statistic is $T = mnS_d^2/(m + n)^2$. For significance at the 5% level, T must exceed 0.461 in a two-tail test and for significance at the 1% level a critical value of 0.743 must be exceeded.

Example 5.13

The problem. Perform the Cramer–Von Mises test for the data sets used in Example 5.12.

Formulation and assumptions. We use the differences in the last column of Table 5.8 to form the statistic T.

Procedure. From the last column of Table 5.8 we find the sum of squares of the differences is 2.002. Since $m = 12$, $n = 16$ we have $T = 12 \times 16 \times 2.002/(28 \times 28) = 0.490$.

Conclusion. Since $T = 0.490$ exceeds the critical value 0.461 we conclude that the population cumulative distribution functions differ for at least some x.

Comments. The Cramer–Von Mises test is not only often more powerful than the Smirnov test, but it is also easy to use because of the good approximation that obviates the need for tables, although more detailed and accurate tables do exist for some m, n. The only labour additional to that for the Smirnov test is calculation of the sums of squares of differences – easy on a pocket calculator or with a computer routine.

5.6 FIELDS OF APPLICATION

It does not require a vivid imagination to think of realistic situations where one might wish to compare medians or means of two populations (i.e. look for location or 'treatment' differences) on the basis of independent samples. Here are a few relevant situations.

Medicine

If we wish to compare the efficacy of two drugs for reducing blood pressure, independent samples may be almost essential because of 'interaction' or 'hangover' if each were tried on the same patients; it might be quite inappropriate for these (and perhaps also for ethical) reasons to give both drugs to any one person, even after a considerable time lapse.

Sociology

To explore the possibility that town and country children may attain different levels of physical fitness, samples of each might be scored in a fitness test and the results compared nonparametrically.

Mineral exploration

A mining company has options to develop two sites but only wishes to develop one. Sample test borings are taken on each and the percentage of the mineral of interest contained in each boring is determined; this information can be used to test for population differences in mean or median levels of mineral; clearly if there is evidence of one site being the richer the company will want to develop it (assuming development costs and other factors are comparable for the two sites).

In most applications tests for equality of variance or Smirnov or Cramer–Von Mises tests for identical population distributions may also be relevant.

EXERCISES

5.1   Verify that in Example 5.1 with samples of 4 and 5, rank sums of 30, 29, 28 for the sample of 4 have respectively probabilities 1/126. 1/126, 2/126.

5.2   In order to compare two different keyboard layouts for a pocket calculator it is designing, a company divides 21 volunteers from their staff randomly into groups of 10 (A) and 11 (B). Each group is asked to carry out the same standard series of calculations, group A using the first type of keyboard, group B the second. Total times (in minutes) for each individual to carry out the complete series are as follows:

$$\begin{array}{llllllllll} \text{Group A} & 23 & 18 & 17 & 25 & 22 & 19 & 31 & 24 & 28 & 32 \\ \text{Group B} & 24 & 28 & 32 & 28 & 41 & 27 & 35 & 34 & 27 & 35 & 33 \end{array}$$

Use the Wilcoxon–Mann–Whitney procedure to obtain a nominal 95% confidence interval for median time difference for the two layouts. Is there evidence that one layout is preferable?

5.3   An alloy is composed of zinc, copper and tin in certain proportions. It may be made at one of two temperatures, denoted by H (for higher) and L (for lower). We wish to know if one temperature produces an alloy which is on the whole harder than the other. We make 9 batches at L and 7 batches at H. To arrange them in order of hardness all specimens are scraped against one another to see which makes the deeper scratch (a deeper scratch indicates a softer specimen). On this basis the specimens are ranked 1 (softest) to 16 (hardest) with the following results.

Temperature	H	L	H	H	H	L	H	L	L	H	H	L	L	L	L	L
Rank	1	2	3	4	5	6	7	8	9	10	11	12	13	14	15	16

Is there sufficient evidence to reject the hypothesis that hardness is unaffected by temperature? State any assumptions needed for validity of the test you use.

5.4   Hotpot Stoves have a standard method of insulating ovens. To test its effectiveness they take random samples from the production line and heat the ovens selected to 400°C, noting the time taken to cool to 350°C after the power is switched off. For a random samples of 8 ovens the times (in minutes) are:

$$15.7 \quad 14.8 \quad 14.2 \quad 16.1 \quad 15.3 \quad 13.9 \quad 17.2 \quad 14.9$$

They decide to explore the use of a cheaper form of insulation and use this on a further sample of 9 ovens and find the times taken for the same temperature drop are:

$$13.7 \quad 14.1 \quad 14.7 \quad 15.4 \quad 15.6 \quad 14.4 \quad 12.9 \quad 15.1 \quad 14.0$$

Are the firm justified in asserting they have no firm evidence of a different rate of heat loss? Obtain a 95% confidence interval for the difference in median rate of heat loss (a) without a making a normality assumption, and (b) with a normality assumption. Comment critically on any difference in your conclusions.

5.5   A psychologist is interested to know whether men or women are more upset by delays in being admitted to hospital for routine surgery. An anxiety index is devised to measure the degree of anxiety shown by patients one week before their scheduled admission. An index measurement is taken for 17 men and 23 women. These are ranked 1 to 40 on a scale of increasing anxiety. The sum of the ranks for the 17 men is 428. Is there evidence against the null hypothesis that anxiety is independent of sex? If there is, which sex appears to show the greater anxiety?

5.6   A psychologist notes the total times (in seconds) needed to perform a series of simple manual tasks for each of 7 children regarded as normal and for each of 8 children regarded as mentally retarded. The times are as follows:

Normal children	204	218	197	183	227	233	191	
Retarded children	243	228	261	202	343	242	220	239

Use a Smirnov test to determine whether the psychologist is justified in asserting that these samples are likely to be from different populations. Do you consider a one- or two-tail test appropriate? Perform also a Cramer–Von Mises test. Does it lead you to a different conclusion? Do you think the psychologist should have tested more specific aspects of any difference? If so, perform appropriate tests.

5.7   Apply the Smirnov test for different population distributions to the oven cooling data in Exercise 5.4.

5.8   Suppose we were given the values for response times in LVF and RVF given in Table 4.1 but the information that they are paired was omitted. In these circumstances we might analyse them as independent samples. What would we then conclude about any median difference between responses in the two fields? Does your conclusion agree with that found in Example 4.1? If not, why not?

5.9   The number of words in the first complete sentence on each of ten pages selected at random is counted in each of the books by Conover (1980) and Bradley (1968). The results were:

Conover	21	20	17	25	29	21	32	18	32	31
Bradley	45	14	13	31	35	20	58	41	64	25

Perform appropriate tests to determine whether there is evidence that in these books:

(i) sentence lengths show a location difference;
(ii) the variance of sentence lengths differs between authors;
(iii) the distributions of sentence lengths differ in an unspecified way;
(iv) the sentence lengths for either author are not normally distributed.

5.10 Lindsey, Herzberg and Watts (1987) give data for widths of first joint of the second tarsus for two species of the insect *Chaetocnema*. Do these data indicate population differences between the width distributions for the two species?

Species I   131 134 137 127 128 118 134 129 131 115
Species II   107 122 144 131 108 118 122 127 125 124

5.11 Carter and Hubert (1985) present data for percentage decreases in blood sugar over 1 hour periods for rabbits given two different doses of a drug. Is there evidence of a response difference between dose levels?

Dose 1	0.21	− 16.20	− 10.10	− 8.67	− 11.13
Dose 2	1.59	2.66	− 6.27	− 2.32	− 10.87
Dose 1	1.96	− 10.19	− 15.87	− 12.81	
Dose 2	7.23	− 3.76	3.02	15.01	

5.12 The journal *Biometrics* published data on the numbers of completed months between receipt of a manuscript for publication and first reply to the authors for each of the years 1979 and 1983. The data are summarized below. Is there evidence of a difference in average waiting time between 1979 and 1983?

Completed months	0	1	2	3	4	5	> 6
No. authors 1979	26	28	34	48	21	22	34
1983	28	27	42	44	17	6	16

5.13 Hill and Padmanabhan (1984) give body weights (g) of diabetic and normal mice. Is there evidence of a significant difference in mean body weight? Obtain the Hodges–Lehmann estimator of the difference together with a 95% confidence interval. Compare the interval with that based on the *t*-distribution.

Diabetic   42 44 38 52 48 46 34 44 38
Normal   { 34 43 35 33 34 26 30 31 31 27 28 27
         { 30 37 38 32 36 32 32 38 42 36 44 33 38

5.14 The data below give information on the numbers of words with various numbers of letters in 200-word sample passages from the Presidential

Addresses to the Royal Statistical Society by W. F. Bodmer (1985) and J. Durbin (1987). Is there acceptable evidence of a difference in average length of words used by the two presidents?

Numbers of letters	1–3	4–6	7–9	10 or more
Bodmer	91	61	24	24
Durbin	87	49	32	32

6

Three or more samples

6.1 POSSIBLE EXTENSIONS

For parametric methods based on normal distribution theory the extension from one or two to several samples shifts emphasis from t-tests and the t-distribution to the analysis of variance and the F-distribution (related because t^2 with v degrees of freedom has an F-distribution with $1, v$ degrees of freedom). Wilcoxon type tests are nonparametric analogues of t-based procedures for the one- or two-sample case. Extending to three or more samples, we find relevant nonparametric techniques closely parallel analysis of variance methods – particularly in computational aspects. The test procedures, especially for moderately large samples, will involve normal approximations and are often based on the F-distribution or the chi-squared distribution (see Section A5).

Readers familiar with design and analysis of experiments will know the term **one-way** analysis of variance, which is relevant to several independent samples. In designed experiments the randomized block design is commonly used as a logical extension to matched pair samples. In this chapter we consider both these extensions in a nonparametric context, dealing first with three or more independent samples.

6.2 LOCATION TESTS FOR INDEPENDENT SAMPLES

The median test introduced in Section 5.1.1 extends to more than two independent samples but we defer consideration of this extension to Section 8.2.4 where it appears as a special case of a more general test.

6.2.1 The Kruskal–Wallis and related tests

A test proposed by Kruskal and Wallis (1952) is a direct extension of the Wilcoxon–Mann–Whitney test expressed in the Wilcoxon formulation.

We test H_0: all samples are from the same distribution, against H_1: at least one sample is from a distribution with a different location.

As in the two-sample case a shift in location is often spoken of as a **treatment effect** in experimental terms.

Suppose we have t random samples and that the ith sample ($i = 1, 2, \ldots, t$) is of size n_i. We denote the individual observations in the ith sample by

$$x_{i1}, x_{i2}, \ldots, x_{in_i}$$

where x_{ij} is the jth observation in the ith sample. The total number of observations is $N = \sum_i n_i$.

We rank all N observations from smallest (rank 1) to largest (rank N). Tied observations are given their mid-rank. Denote by r_{ij} the rank allotted to the observation x_{ij}. Let s_i be the sum of the ranks in the ith sample (analogues of the Wilcoxon s_m, s_n). We compute $S_t^2 = \sum_i (s_i^2/n_i)$ and $S_r^2 = \sum_{i,j} r_{ij}^2$. With no ties $S_r^2 = N(N+1)(2N+1)/6$. The reader familiar with the analysis of variance will recognize these as uncorrected **treatment** and **total** sums of squares for ranks. From each we subtract appropriate correction for the mean; here $C = \frac{1}{4}n(n+1)^2$. The test statistic is

$$T = \frac{(N-1)[S_t^2 - C]}{(S_r^2 - C)} \qquad (6.1)$$

With no ties T simplifies to

$$T = 12S_t^2/[N(N+1)] - 3(N+1) \qquad (6.2)$$

With only a few ties the difference between (6.1) and (6.2) is small and if calculations are done on a pocket calculator the simpler (6.2) will suffice. For a computer program (6.1) is preferable to cover all cases. Tables giving critical values for small sample sizes are available (see e.g. Neave, 1981, pp. 32–4), but T has approximately a chi-squared distribution with $t - 1$ degrees of freedom for moderate or large N (see Section A5 if there are no treatment differences).

Example 6.1

The problem. As well as the statistics and general books in Example 5.2, I have another shelf containing 135 travel and nature books. I counted numbers of pages in a random sample of 8 of these. The numbers are reproduced, together with those for the samples of general and statistics books introduced in Example 5.2, in Table 6.1. Use the Kruskal–Wallis test to examine the validity of the hypothesis that these may be samples from the same population.

Formulation and assumptions. The alternative hypothesis is that the samples come from populations exhibiting at least one location difference. With $N = 36$ and sample sizes 8, 16, 12 the chi-squared approximation is reasonable.

Procedure. Table 6.1 gives the data and the combined samples rank for each observation. For convenience in allocating ranks the observations are given in ascending order in each sample.

Here $s_1 = 6 + 7 + 15 + 20 + 25 + 26 + 34 + 36 = 169$; similarly $s_2 = 227$ and $s_3 = 270$. Also $n_1 = 8, n_2 = 16, n_3 = 12$ and $N = 8 + 16 + 12 = 36$. We find $S_t^2 = (169^2/8) + (227^2/16) + (270^2/12) = 12\,865.69$. There are no ties so we use the form (6.2), whence

$$T = 12 \times 12\,865.69/(36 \times 37) - 3 \times 37 = 4.91$$

Table 6.1 Page numbers and ranks in three independent book samples

Travel:											
Pages	93	98	216	249	301	319	731	910			
Ranks	6	7	15	20	25	26	34	36			
General:											
Pages	29	39	60	78	82	112	125	170	192	224	
Ranks	1	2	3	4	5	8	9	13	14	16	
Pages	263	275	276	286	369	756					
Ranks	21	22	23	24	27	35					
Statistics:											
Pages	126	142	156	228	245	246	370	419	433	454	478	503
Ranks	10	11	12	17	18	19	28	29	30	31	32	33

Tables for the chi-squared distribution (see e.g. Neave, 1981, p. 21) show that the critical value that must be exceeded for significance at the 5% level with 2 degrees of freedom in a one-tail test is 5.991.

Conclusion. Since $T = 4.91$, less than the critical value of 5.991, we do not reject the null hypothesis that the samples all come from the same population.

Comments. 1. If we relax our usual criterion and allow significance at the 10% level, the critical chi-squared value is 4.605 so we could reject the null hypothesis at this less than usually stringent level.

2. In Example 5.1 we found a significant difference at the 5% level between general and statistics books. It therefore seems an anomaly that introducing a different class of book (travel) has apparently removed this significant difference. This sort of anomaly sometimes happens when one incorporates a portion of data into a larger analysis, or alternatively analyses separately only a portion of the available data; it is generally (and rightly) regarded as bad statistical practice to hive off sections of data in a way that happens to suit us. We take up the question of looking at particular aspects of data when we discuss multiple comparison tests in Section 6.4.2.

3. We did not establish significance at the conventional 5% level. Would the standard parametric analysis of variance serve us better? We find the appropriate variance ratio for an analysis of variance of the original data gives an F-value not even significant at the 10% level. Possible reasons are non-normality, heterogeneity of variance or presence of 'outliers'. Certainly, although there are some short to average-length travel books, there are also two very long ones – 731 and 910 pages. We investigate possible heterogeneity of variance in Section 6.3; however, by using ranks the Kruskal–Wallis test

shows, in general, more robustness against such heterogeneity than the conventional analysis of variance.

4. A check that we have correct values of the s_i is that they should sum to $\frac{1}{2}N(N+1)$. In this example $N = 36$.

6.2.2 A normal scores test

An obvious modification to the Kruskal–Wallis test is to use normal scores in place of ranks. We illustrate this using van der Waerden scores. As in the two-sample case we replace rank r by the $r/(N+1)$th quantile for a standard normal distribution. We use quantiles corresponding to mid-ranks for ties, when the following test becomes an approximation that makes little difference for moderate numbers of ties.

The test statistic T is an analogue of that for the Kruskal–Wallis test; simpler because there is no correction term since the mean of the van der Waerden scores is zero if there are no ties (still approximately so with a few ties).

Denote by a_{ij} the van der Waerden score corresponding to the rank r_{ij} assigned to the original observation x_{ij}. Let a_i be the sum of the a_{ij} for the ith sample. We calculate $A_t^2 = \sum_i(a_i^2/n_i)$; the analogue of S_t^2 in the Kruskal–Wallis test. We also calculate $A_r^2 = \sum_{i,j}a_{ij}^2$. The test statistic is

$$T = (N-1)A_t^2/A_r^2 \tag{6.3}$$

There is a close analogy between (6.1) and (6.3), remembering that the mean of the normal scores is zero so $C = 0$ in (6.3). Again the test is performed by comparing T with the critical chi-squared value with $t - 1$ degrees of freedom.

Example 6.2

The problem. Repeat the analysis in Example 6.1 using van der Waerden scores in place of ranks.

Formulation and assumptions. We rank all observations as in Example 6.1 then obtain appropriate quantiles from tables of the standard normal distribution and form the statistic (6.3).

Procedure. Table 6.2 reproduces Table 6.1 and introduces an additional line of van der Waerden scores, each given to two decimal places. For example, for the third observation on travel books, $r_{1.3} = 15$, so the van der Waerden score is the $15/37 = 0.4054$th normal quantile. From tables of the standard normal distribution we find $\Pr(Z < -0.24) = 0.4052$. Thus -0.24 is (to two decimal places) the appropriate van der Waerden score.

We need the sum of scores for each sample. For the first sample it is

$$a_1 = -0.99 - 0.88 - 0.24 + 0.10 + 0.46 + 0.53 + 1.40 + 1.92 = 2.30$$

Table 6.2 Pages, ranks and scores in three independent book samples

Travel:

Pages	93	98	216	249	301	319	731	910	
Ranks	6	7	15	20	25	26	34	36	
Scores	-0.99	-0.88	-0.24	0.10	0.46	0.53	1.40	1.92	

General:

Pages	29	39	60	78	82	112	125	170	
Ranks	1	2	3	4	5	8	9	13	
Scores	-1.92	-1.61	-1.40	-1.24	-1.10	-0.79	-0.70	-0.38	
Pages	192	224	263	275	276	286	369	756	
Ranks	14	16	21	22	23	24	27	35	
Scores	-0.31	-0.17	0.17	0.24	0.31	0.38	0.61	1.61	

Statistics:

Pages	126	142	156	228	245	246	370	419	433
Ranks	10	11	12	17	18	19	28	29	30
Scores	-0.61	-0.53	-0.46	-0.10	-0.03	0.03	0.70	0.79	0.88
Pages	454	478	503						
Ranks	31	32	33						
Scores	0.99	1.10	1.24						

Similarly $a_2 = -6.30$ and $a_3 = 4.00$. Note that $a_1 + a_2 + a_3 = 0$ as there are no ties. (It would be approximately zero even if there were a few ties.)

We find $A_t = 2.30^2/8 + (-6.30)^2/16 + 4.00^2/12 = 4.4752$. The sum of squares of all 36 scores is 30.1114, whence $T = 35 \times 4.4752/30.1144 = 5.20$. The test then proceeds as in Example 6.1. The T value again just fails to reach significance at the 5% level.

Conclusion. Clearly, as in Example 6.1, the probability of getting as high or a higher value of T when H_0 is true is less than 0.10 but greater than 0.05, so we would have to relax our usual 5% criterion to reject the null hypothesis.

Comments. The slightly higher value of the test statistic in Example 6.2 is consistent with a higher efficiency than that based on ranks. This is in line with the known higher Pitman efficiency of normal scores tests for many families of distributions. However, not too much weight should be put on this apparent improvement as the approximations to the chi-squared distribution are not necessarily equally good.

6.2.3 The Jonckheere test

The Kruskal–Wallis test is an omnibus test of differences between locations. If treatments represent steadily increasing doses we may want to test H_0: all μ_i

are equal (i.e. no treatment effect), against $H_1: \mu_1 \leqslant \mu_2 \leqslant \cdots \leqslant \mu_t$, where at least one inequality is strict. We may also consider H_1 with all inequality signs reversed if this is appropriate. For such ordered H_1 a test proposed by Jonckheere (1954) is more powerful than the Kruskal–Wallis test. The test is described by Daniel (1978, Section 6.3) and in a very detailed manner by Leach (1979, Section 5.1).

6.3 TESTS FOR HETEROGENEITY OF VARIANCE FOR INDEPENDENT SAMPLES

The squared rank test given in Section 5.4.2 extends easily to several independent samples. Again, the computational procedure bears a striking similarity to some used in the analysis of variance. As in the two-sample case, we first replace the observations by the absolute deviations from the sample mean for each sample. These deviations are ranked across all samples and the squares of ranks are obtained. The analysis is based on these squared ranks. The statistic T is identical in form to (6.1) except that now s_i is the sum of the squared rank deviations for sample i, and the correction C is now the square of the sum of the squared ranks divided by N; symbolically, denoting a squared rank by r_{ij}^2, $C = [\sum_{i,j} r_{ij}^2]^2/N$. We replace S_r^2 in (6.1) by the sum of squares of the squared ranks, i.e. $S_r^2 = \sum_{i,j} r_{ij}^4$. For reasonably large samples under H_0: all population variances equal, T has a chi-squared distribution with $t - 1$ degrees of freedom.

If there are no ties the sum of squared ranks simplifies to $N(N + 1) \times (2N + 1)/6$. Also, in this case the denominator of T reduces to

$$S_r^2 - C = (N - 1)N(N + 1)(2N + 1)(8N + 11)/180$$

Example 6.3

The problem. For the data in Example 6.1 use squared ranks to test H_0: all population variances are equal against H_1: there is heterogeneity of variance.

Formulation and assumptions. We first obtain the absolute deviation $|x_{ij} - \bar{x}_i|$ of each observation from its sample mean $\bar{x}_i$. These deviations are then ranked over all samples combined; the statistic T is calculated and compared with the relevant chi-squared table entry.

Procedure. The means of the original observations (numbers of pages) are:

Travel books   364.6
General books   208.5
Statistics books   316.7

Table 6.3 gives the absolute deviations (in the same order as the original observations in Table 6.1); e.g. the first entry for general books is

Table 6.3 Deviations, ranks and squared ranks for page numbers

Travel:								
Deviations	271.6	266.6	148.6	115.6	63.6	45.6	366.4	545.4
Rank	33	32	23	17	7	4	34	35
Squared rank	1089	1024	529	289	49	16	1156	1225
General:								
Deviations	179.5	169.5	148.5	130.5	126.5	96.5	83.5	38.5
Rank	29	27	22	20	19	15	13	3
Squared rank	841	729	484	400	361	225	169	9
Deviations	16.5	15.5	54.5	66.5	67.5	77.5	160.5	547.5
Rank	2	1	6	8	9	12	24	36
Squared rank	4	1	36	64	81	144	576	1296
Statistics:								
Deviations	190.7	174.7	160.7	88.7	71.7	70.7	53.3	102.3
Rank	31	28	25	14	11	10	5	16
Squared rank	961	784	625	196	121	100	25	256
Deviations	116.3	137.3	161.3	186.3				
Rank	18	21	26	30				
Squared rank	324	441	676	900				

$|29 - 208.5| = 179.5$. Below each absolute deviation we give its overall rank and below that the square of that rank used in forming T. In practice, to reduce chances of error when the analysis is not programmed for a computer, it may help to rearrange absolute deviations within each sample in ascending order (as we did for the original observations in Table 6.1).

The sums of squared ranks for each sample are respectively 5377, 5420 and 5409, whence

$$S_t^2 = (5377)^2/8 + (5420)^2/16 + (5409)^2/12 = 7\,888\,147.88$$

There are no ties; the sum of all squared ranks is $(36 \times 37 \times 73)/6 = 16\,206$, whence $C = (16\,206)^2/36 = 7\,295\,401$. Also

$$S_r^2 = 35 \times 36 \times 37 \times 73 \times 299/180 = 5\,653\,193$$

whence

$$T = 35(7\,888\,147.88 - 7\,295\,401)/5\,653\,193 = 3.67$$

Comparison with the chi-squared distribution with 2 degrees of freedom indicates that this is not significant even at the 10% level.

Conclusion. We do not reject the null hypothesis that the samples come from populations with the same variance.

Comments. Note that in carrying out the squared ranked test we align the populations for location. This does not affect the variance. For the original data, of course, there may be differences of location whether or not we accept H_0 with the squared rank test.

We could use a squared normal scores test for homogeneity of variance. One follows the procedure in Example 6.3 to the stage where deviations are ranked, but then replaces those ranks by their van der Waerden scores which are then squared. The analysis proceeds on lines that closely parallel those just given for the squared ranks. For the data in Example 6.3 this procedure produces almost the same value of T (3.69 instead of 3.67).

The tests in this section may be insensitive to some forms of heterogeneity of variance; for example, if these are associated also with differences in skewness. An alternative measure of centrality would be to use sample medians in place of sample means as a basis for calculating absolute deviations.

6.4 FURTHER TESTS FOR SEVERAL INDEPENDENT SAMPLES

6.4.1 Kolmogorov–Smirnov type tests

Work has been done on testing H_0: all samples from identical distributions, against H_1: at least one distribution differs; also on testing a null hypothesis that all samples come from the same specified distribution. Little is known about the distribution of the statistics used for such tests. A few tables exist for the special case of small samples of equal size. Simulation studies provide the best hope of extending such tables to unequal sample sizes with the possibility that asymptotic theory may allow approximate tests for large samples. We do not pursue this matter. Conover (1980, Section 6.4) deals with some equal sample size cases.

6.4.2 Multiple comparisons

Readers familiar with analysis of experimental results by parametric methods will know that the parametric analogue of the Kruskal–Wallis type analysis is usually only a preliminary to a more detailed study of differences between individual samples or groups of samples. These multiple comparisons, as they are called, require some care in selection of appropriate tests or estimation procedures in the parametric case. Nonparametric analogues to many of these tests appear in the literature but it is fair to say that they are more restricted than their parametric counterparts partly because, in the absence of exact tables, we are dependent on large-sample approximations for establishing significance.

We confine our attention to tests equivalent to what are known as **least significant differences**. Logically they can only be justified if applied to pre-selected comparisons when an overall Kruskal–Wallis or equivalent normal scores test indicates significance, and the significance level used in these tests should be at no less stringent a level than that in the overall test.

For the Kruskal–Wallis test the criteria for saying that the medians differ for two samples, say, the ith and the jth sample, are

1. that the overall test indicates significance and
2. that if s_i', s_j' are the mean ranks for those samples (i.e. the total ranks (scores) for the sample divided by n_i, n_j respectively; $s_i' = s_i/n_i$), then

$$|s_i' - s_j'| > t_{n-t,\alpha}\sqrt{[(S_r^2 - C)(N - 1 - T)}$$
$$\times (n_1 + n_j)/\{n_1 n_j(N - t)(N - 1)\}] \tag{6.4}$$

where $t_{n-t,\alpha}$ is the value of t required for significance at the $100\alpha\%$ level in a t-test with $n - t$ degrees of freedom, and the other quantities are as defined in Section 6.2.1.

An analogous criterion holds for van der Waerden scores with sample rank means replaced by sample score means, and $S_r^2 - C$ replaced by A_r^2 as defined in (6.3), which also provides the relevant T for this case.

Many statisticians agree with Pearce (1965, pp. 21–2) that it is not unreasonable to compare a sub-set of all samples if that sub-set is selected before the data are obtained (or at least before they are inspected) using formulae like (6.4) if the experimental logic suggests it is reasonable to do so in the sense that one could sensibly expect the specified pairs to show differences in location.

In the example on books there may be no prima facie case to suggest that general books might be shorter than statistics books unless one had some evidence that statistics authors tended to be more verbose than most. I might, however, have a suspicion that statistics books tend to be longer if I bulk-weighed all my statistics books and got an average weight by dividing total weight by the number of books and also did the same thing for my general books and found their average weight were lower. Of course, a different average number of pages is not the only possible explanation for differences in mean weight. This could be due to several causes. The statistics collection may include more hardcover books, they may be printed on thicker paper, they may have larger page sizes. If I have a general impression that none of these factors are particularly influential in accounting for weight differences then it would seem reasonable to test if the mean number of pages differed.

Example 6.4

The problem. After a weighing experiment like that just described I feel it likely that the median page numbers may be higher for my statistics books than for my general books. On these grounds I rate this a nominated comparison. Test at the 5% level by a least significant difference test whether I would be justified in believing these samples have different medians.

Formulation and assumptions. We may use either the information from the Kruskal–Wallis test (Example 6.1) or the normal score test (Example 6.2) in (6.4). For illustrative purposes we use both. Note that a 5% test is more stringent than the 10% level needed for significance in the overall test.

Procedure. 1. From Example 6.1 we easily find that the relevant quantities in (6.4) are $N = 36, t = 3, n_2 = 16, n_3 = 12, s_2' = 227/16 = 14.19, s_3' = 227/12 = 22.5, \quad S_r^2 - C = 36 \times 37 \times 73/6 - 36 \times 37 \times 37/4 = 3885, \quad T = 4.91.$ From t-distribution tables (see, e.g. Neave, 1981) we find $t_{33,0.05} = 2.03$.
   Thus the left-hand side of (6.4) has the value $s_3' - s_2' = 8.26$ while the right-hand side is
$$2.03\sqrt{[\{3885(36 - 1 - 4.91)(16 + 12)\}/(16 \times 12 \times 33 \times 35)]} = 7.80$$

2. Similar calculations based on relevant values derived from the normal score analysis in Example 6.2 are left as an exercise for the reader. In this case the left-hand side has the value 0.727 (be careful with the negative sign for one of the score means!). The right-hand side is 0.683.

Conclusion. In both procedures 1 and 2 the left-hand side of the inequality exceeds the right, so we conclude there is a location difference for these two samples.

Comments. This finding is consistent with that in Example 5.1 where we analysed data for these two samples alone, ignoring the third sample of travel books. In general, an analysis based on additional data will tend to be more powerful than one on less data (providing the relevant method is used), but that is not apparent in this case.

We have, for illustrative purposes, analysed book data in a single-sample case, a two independent sample situation and a three independent sample case. The three-sample analyses were direct extensions of the corresponding two-sample ones and given all the data it is most appropriate to make inferences about the two samples from a relevant three-sample analysis; e.g. given all three samples initially, we would not have carried out the analyses in the previous chapter. It would certainly be a mistake to carry out all pairwise analyses between three samples by the method in the last chapter. This would be equivalent to comparing all treatments pairwise by something very like the least significant difference method of this section. Such tests are not independent and this affects significance levels. In particular, a test of data for a large number of treatments comparing the two means that exhibit the greatest difference (selected after the data have been inspected) using a least significant difference test (parametric or nonparametric) will nearly always give a wrong significance level. The level used may well be nearer 90% than 5%; i.e. the probability of an error of the first kind will be nearer 90% than 5%.

Multiple comparison tests have been widely extended in scope in recent years for both parametric and nonparametric methods. Leach (1979, Section 4.2) suggests a rather more conservative approach than that described here.

6.5 LOCATION COMPARISONS FOR RELATED SAMPLES

If we have blocks of two units and allocate two treatments to units in each block at random we have the matched pair situation discussed in Chapter 4. This is also the simplest case of a randomized block design. One reason we used matched pairs was to allow us to make comparisons within blocks (simply by analysing differences between treatment responses). We did this because we expected results for each member of the pair to be rather more homogeneous under the null hypothesis of no treatment effect than they would be if treatments were applied randomly to large sets of very heterogeneous units.

If we wish to compare several (say t) treatments we replace our homogeneous pairs by blocks of t units, blocks being chosen so that the units in each block are as homogeneous as possible. For example, in an animal experiment, to compare the effect of five different feed regimes on pigs, ideally our first block might be five pigs from the same litter, our next five pigs from a second litter, the next five from a third litter, and so on; in each litter the five diets are allocated, one to each pig, chosen at random.

To compare three different cake recipes (treatments) we might make batches of each recipe and divide each batch into four samples and cook one sample of each batch in four different ovens (each oven is a block). We do this because ovens may operate at slightly different temperatures and have varying heat efficiency however much we try to control such factors. If we are going to assess the merits of the cake recipe by asking experts to rank them in order of preference for taste, then comparison of mixtures cooked in the same oven is desirable because, although all mixtures may produce poor cakes in some ovens and all rather nice ones from other ovens, one hopes that the relative orderings for products from different ovens may be reasonably consistent. We have so far not considered randomization within ovens. If all mixtures are cooked at the same time it might be wise to allocate the three different mixtures to shelf positions at random as these may affect the cakes produced, or if the cakes are cooked one after the other in each oven the time order could sensibly be randomized separately for each oven as the end product might be affected by the time the mixture stands before cooking, or there may be some carry-over effect on flavour dependent on which cake was cooked first. These are matters reflecting principles of good experimental design and are relevant to all methods of analysis – parametric or nonparametric.

6.5.1 A generalization of the sign test

In Section 4.1.2 we applied the sign test to a matched pair situation. Effectively what we did there was allocate to the treatment difference a plus or a minus sign. We could have done an exactly equivalent analysis by ranking treatments within each block. If we gave rank 1 to the lower response and rank 2 to the higher response unit in each block and then sum all ranks allocated to each treatment we may deduce the number of plus and minus signs, and the difference between the numbers of each, as the following simple example shows. Suppose the data in Table 6.4 represent observations on 9 matched pairs numbered 1 to 9.

Table 6.4 Observations in matched pairs

Pair	1	2	3	4	5	6	7	8	9
Treatment A	17	11	15	14	22	41	7	2	8
Treatment B	15	9	18	11	17	43	5	4	3

A quick visual examination indicates that the difference treatment A – treatment B gives rise to 6 positive and 3 negative values, these being the relevant numbers of plus and minus signs for a sign test.

If we rank the observations in increasing order in each pair we get Table 6.5.

Table 6.5 Rank order of responses in Table 6.4

Pair	1	2	3	4	5	6	7	8	9
Treatment A	2	2	1	2	2	1	2	1	2
Treatment B	1	1	2	1	1	2	1	2	1

It is immediately apparent that the number of positive signs is equal to the number of 2's allocated to treatment A (or the number of 1's allocated to treatment B). This result is perfectly general for any number of pairs so the rankings for either treatment immediately lead to the sign test statistic. Also the difference between the rank sums for each treatment equals the difference between the number of plus and minus signs in a sign test.

We generalize this idea to more than two treatments when we have a randomized blocks design for which the number of units in a block equals the number of treatments.

In the parametric analysis of variance of continuous observations we 'remove' block difference as a source of variability before making comparisons in the relevant analysis (but do not worry about how this is done if you are unfamiliar with the analysis of variance). Friedman (1937) proposed the test described below which automatically removes block differences by the simple expedient of replacing observations by their ranks, performing the operation separately within each block. The test is also applicable to data given only as ranks within each block (see Exercise 6.4). In this latter circumstance it becomes a test for consistency of ranking rather than one for a location parameter. The test in this context was first developed by M. G. Kendall (see Kendall, 1962) and we discuss it further in Section 7.4.

Suppose we have t treatments each applied to one of the t units in each of b blocks in a randomized block design. We denote by x_{ij} the response (observation) for treatment i in block j. Here i runs from 1 to t and j runs from 1 to b. In the Friedman test we replace the observations in each block by ranks 1 to t. This ranking is carried out separately for each block. The sum of the ranks is then obtained for each treatment. We denote these by s_i, $i = 1, 2, \ldots, t$. We require also the sum of squares of all ranks (equivalent to the uncorrected total sum of squares in an analysis of variance). If r_{ij} denotes the rank (or appropriate mid-rank if there are ties within any block) of x_{ij} this uncorrected (sometimes called raw) total sum of squares is $S_r^2 = \sum_{i,j} r_{ij}^2$. With no ties $S_r^2 = bt(t+1)(2t+1)/6$. We also require an uncorrected sum of squares between treatment ranks which is, by analogy with the parametric analysis of variance, $S_t^2 = \sum_i (s_i^2)/b$. The correction factor, analogous to that for the Kruskal–Wallis test, is $C = \frac{1}{4}bt(t+1)^2$. A form of the Friedman statistic commonly used is

$$T_1 = b(t-1)(S_t^2 - C)/(S_r^2 - C) \qquad (6.5)$$

For b, t not too small, T_1 has approximately a chi-squared distribution with $t-1$ degrees of freedom. For values of $t \leqslant 6$, Neave (1981, p. 34), gives tables of critical values of T_1.

Iman and Davenport (1980) suggested that a better approximation is given by

$$T = (b-1)(S_t^2 - C)/(S_r^2 - S_t^2) \qquad (6.6)$$

which, under the null hypothesis of no treatment difference, has approximately an F-distribution with $t-1$ and $(b-1)(t-1)$ degrees of freedom.

Occasionally the denominator of this statistic is zero. This occurs if the ranking is identical in all blocks. In this case the result is significant at the $100(1/t!)^{b-1}\%$ significance level. We use (6.6) in preference to (6.5) in the following example.

Example 6.5

The problem. Pearce (1965, p. 37) quotes results of a greenhouse experiment carried out by Janet I. Sprent (unpublished). The data, given in Table 6.6, are numbers of nodes to first initiated flower summed over four plants in each experimental unit (pot) for the pea variety Greenfeast for six treatments – one an untreated control, the others various named growth substances. There were four blocks (the grouping allowed for different light intensities and temperature gradients depending on proximity to the greenhouse glass, blocks being arranged to make these conditions as like as possible for all units – pots of four plants – in any one block). Apply the Friedman test for differences between treatments in node of flower initiation.

Table 6.6 Nodes to first flower, total for four plants

Block	I	II	III	IV
Treatment				
Control	60	62	61	60
Giberellic acid	65	65	68	65
Kinetin	63	61	61	60
Indole acetic acid	64	67	63	61
Adenine sulphate	62	65	62	64
Maleic hydrazide	61	62	62	65

Formulation and assumptions. We replace the observations by ranks within each treatment and calculate the Friedman statistic for these ranks to test H_0: no difference between treatments, against H_1: at least one treatment has a different location from the others. We test using (6.6) and the F-approximation.

Procedure. The relevant ranks are given in Table 6.7 and for convenience we append a further column giving rank totals, s_i, for each treatment.

Table 6.7 Nodes to first flower, ranks within blocks

Block	I	II	III	IV	Total
Treatment					
Control	1	2.5	1.5	1.5	6.5
Giberellic acid	6	4.5	6	5.5	22
Kinetin	4	1	1.5	1.5	8
Indole acetic acid	5	6	5	3	19
Adenine sulphate	3	4.5	3.5	4	15
Maleic hydrazide	2	2.5	3.5	5.5	13.5

The total rank sum of squares obtained by squaring each rank and adding is $S_r^2 = 361$. The uncorrected treatment sum of squares is

$$S_t^2 = (6.5^2 + 22^2 + 8^2 + 19^2 + 15^2 + 13.5^2)/4 = 339.625$$

The correction, $C = 4 \times 6 \times 7^2/4 = 294$.
Thus T, given by (6.6), is

$$T = 3(339.625 - 294)/(361 - 339.625) = 6.40$$

Critical values of F with 5 and 15 degrees of freedom may be obtained from tables such as Neave (1981, pp. 22–5) and are 2.901 at the 5% level, 4.556 at the 1% level, 7.57 at the 0.1% level.

Conclusion. Since our T value of 6.40 exceeds the critical value 4.556 we reject the null hypothesis of no treatment difference at the 1% level; i.e. the difference is highly significant.

Comments. 1. A parametric analysis of variance of the original data produced an F-value of 4.56 which just indicates significance at the same level. If we use (6.5) for the Friedman test we reach the same basic conclusion, but the large-sample chi-squared approximation does not quite reach significance at the 1% level.

2. Whereas the ordinary analysis of variance introduces a sum of squares to reflect differences between block totals in the original data, there is no such term in the Friedman analysis because the sum of ranks in all blocks is the same, namely $\frac{1}{2}t(t + 1)$. It is easily confirmed that this is 21 for all blocks in Table 6.6. Thus, as already indicated, the Friedman procedure removes block differences by the method of ranking.

It is feasible to replace ranks within each block by corresponding normal scores. Experience with the Friedman method suggests there is likely to be little advantage in such a rescoring. Indeed if there are ties the block differences removed by the ranking process may be reintroduced, though usually not dramatically.

A major advantage of ranking within blocks is that it provides a method that is reasonably robust against many forms of heterogeneity of variance in that it removes any inequalities between blocks (see e.g. Exercise 6.8).

Page (1963) proposed an analogue to Jonckheere's test for ordered alternatives applicable to blocked data. It is described by Daniel (1978, Section 7.3) and by Marascuilo and McSweeney (1977, Section 14.12).

6.5.2 Extension of Wilcoxon type tests

We pointed out that the Friedman test is a natural extension of the sign test. It is logical to seek extensions of Wilcoxon tests. One such test is the so-called **aligned ranks test** where deviations from the block mean for each block are ranked over all blocks. Details are given by Lehmann (1975, pp. 270–3). The method has intuitive appeal but only large-sample approximations are available for testing and there is some doubt whether the test for large samples is more efficient than the Friedman approach using the corresponding approximation. Another approach was suggested by Quade (1979) and is described by Conover (1980, Section 5.8), who reports unpublished studies suggesting the Friedman test is more powerful in all but small samples. The Quade test weights ranks within blocks in a rather complicated manner giving greater weights in blocks that are more variable (as indicated by the range of values within a block). This has little intuitive appeal and it is possible to invent data sets where fairly obvious differences in the original data are not picked up by the Quade test.

A more promising approach is described by Hora and Conover (1984). All observations are ranked simultaneously without regard to treatment or blocks and the usual analysis of variance is carried out on ranks or normal scores based on these. For ranks the procedure is described by Iman, Hora and Conover (1984).

When applied to the data in Example 6.5 the relevant F-statistic for treatment differences has the value 4.96 (compared with 4.56 for a parametric test and 6.40 using (6.6)). The reader familiar with standard analysis of variance may care to verify this result (see Exercise 6.7).

6.5.3 Multiple comparisons with the Friedman test

If the Friedman test indicates overall significance, significance for nominated comparisons may be tested by examining differences between relevant treatment rank totals or means.

For totals, the magnitude of these differences is adjudged significant if they exceed the least significance difference given by

$$t_{(b-1)(t-1),\alpha}\sqrt{[2b(S_R^2 - S_T^2)/(b-1)(t-1)]} \tag{6.7}$$

Example 6.6

The problem. The pea node data in Example 6.5 include a control treatment given no growth substance because the experimenter wished to compare all other treatments with this as a base. Regarding these as nominated comparisons, check whether any exceed the least significant difference.

Formulation and assumptions. For nominated comparisons a test based on (6.7) is relevant.

Procedure. Since $t = 6$, $b = 4$, the t-value has 15 degrees of freedom and from tables we find $t_{15,0.05} = 2.13$. In Example 6.5 we found $S_r^2 = 361$ and $S_t^2 = 339.625$ whence the least significant difference is $2.13\sqrt{[2 \times 4 \times (361 - 339.625)/(3 \times 5)]} = 7.19$.

Conclusion. The sums of ranks for each treatment are given in the last column of Table 6.7. From there we see that the differences

$$\text{Gibberellic acid} - \text{control} = 22 - 6.5 = 15.5$$

$$\text{Indole acetic acid} - \text{control} = 19 - 6.5 = 13.5$$

$$\text{Adenine sulphate} - \text{control} = 15 - 6.5 = 8.5$$

We conclude the node numbers are significantly higher with additions of each of these growth substances.

Comments. 1. The analysis of variance of the original data described by Pearce (1965, pp. 36–8) reaches very similar conclusions but in that case the adenine sulphate – control difference just fails to reach the least significant difference.

2. As the overall test indicated significance at the 1% level, strictly speaking least significant differences should be specified at the same level. This would exclude significance for the adenine sulphate – control contrast.

A more sophisticated approach to multiple comparisons is given by Leach (1979, Section 6.2). A recent paper on this topic is that by Shirley (1987) and an important earlier paper is one by Rosenthal and Ferguson (1965).

6.5.4 Cochran's test for binary responses

The Friedman test is applicable to data arranged in blocks that are at least ordinal and are or may be ranked within blocks. We often meet situations where the response of a unit may be one of two possibilities – win or lose; success or failure; alive or dead; male or female. For analytic purposes these responses are labelled 0 or 1. For example, five members A, B, C, D, E of a mountaineering club may each attempt three rock climbs at each of which they succeed or fail. If a success is recorded as 1, a failure as 0, the outcomes may be summarized as follows:

Member	A	B	C	D	E
Climb no. 1	1	1	0	0	1
Climb no. 2	1	0	0	1	0
Climb no. 3	0	1	1	1	1

Cochran (1950) proposed a test that could be applied to this example to test the hypothesis H_0: all climbs are of equal difficulty, against the alternative, H_1: they vary in difficulty. In conventional terms the 'climbs' are the treatments and the 'climbers' the blocks. If we have t treatments and b blocks with binary (i.e. 0, 1) responses the appropriate test statistic is

$$Q = \frac{t(t-1)\sum_i T_i^2 - (t-1)N^2}{tN - \sum_j B_j^2}$$

where T_i is the total (of 1's and 0's) for treatment i, B_j is the total for block j and N is the grand total.

The exact distribution of Q is difficult to obtain, but for large samples Q has approximately a chi-squared distribution with $t-1$ degrees of freedom.

Although it is not immediately obvious, the test reduces to the McNemar test for two treatments. Indeed, we introduced the McNemar test with an example comparing two climbs in Section 4.2, and we shall give an alternative form of the McNemar test in Section 8.2.5 which is exactly equivalent to using Q above.

Cochran's test is discussed more fully by Conover (1980, Section 4.6) and also by Leach (1979, Section 6.3), Daniel (1978, Section 7.5) and by Marascuilo and McSweeney (1977, Section 7.7).

6.6 FIELDS OF APPLICATION

Parametric analysis of variance of designed experiments had its historical origins in agricultural experiments but soon spread to the life and natural sciences, medicine, industry and more recently to business and the social sciences. The need for nonparametric analogues was stimulated by a realization that data often clearly violated some of the rather stringent assumptions needed to validate normal theory analysis of variance, but more importantly perhaps to provide a tool for analysing ordinal data expressed only by ranks or preferences. Ordinal data arise in a number of situations. Ranking or other simple scores of preference often combine assessments of a number of factors that are given different importance by individuals; it is therefore of considerable practical interest to see if there is still consistency between the way individuals rank the same objects despite the fact that they give different weights to different factors. The first example below illustrates this point.

Preference for washing machines

Consumers' preference for a particular washing machine will reflect their assessment of a number of factors; price, reliability, power consumption, time taken to wash various fabrics, efficiency with which each is cleaned, tendency to damage more delicate fabrics, total load capacity, ease of operation, clarity of operating instructions, etc.

Different people will give different weights to each factor; a farmer's wife who offers bed and breakfast to visiting tourists is likely to rate the ability to wash bed linen highly; a parent with a lot of children, the ability to remove sundry stains from children's clothing. To one who is impecunious, a low price and economy of running may be key factors. Manufacturers of any washing machine know they will not get top ratings on all points; what they are keen to achieve is a high overall rating of their product from a wide range of consumers. If each of a number of people are asked to state preferences (i.e. ranks) for one manufacturer's machine and those of several competitors the manufacturer is keen to know if there is consistency in rankings – whether most people give that machine a preferred rating – or whether there is inconsistency – or a general dislike of that machine. Each consumer is a block, each machine a treatment, in the context of the Friedman test described in Section 6.5.1. The hypothesis under test is H_0: there is no consistency in rankings, against H_1: there is some consistency in rankings. We discuss this type of situation further in Section 7.4.

Literary discrimination

A professor of English may claim that he thinks short stories by a certain writer are excellent, those by a second writer are good, and those by a third inferior. To test his claims and judgement he is given 20 short stories to read on typescripts that do not identify the authors and asked to rank them 1 to 20 (1 for the best, 20 for the worst). In fact 6 are by the author he rates as excellent, and 7 each by the authors he rates as good and inferior. The rankings given by the professor when checked against authors are

> Excellent author: 1, 2, 4, 8, 11, 17
> Good author: 3, 5, 6, 12, 16, 18, 19
> Inferior author: 7, 9, 10, 13, 14, 15, 20

Does this classification justify his claim of discriminatory ability? A Kruskal–Wallis test could be used to test the hypothesis that the ranks indicate these could all come from the same population, i.e. there is insufficient evidence to indicate that he has discriminatory power. See Exercise 6.2.

Assimilation and recall

A list of 12 names is read out to students. It contains, in random order, 4 names of well-known sporting personalities, 4 of national and international political figures, and 4 prominent in local community affairs. The students are later asked how many names they can recall and a record is made of how many names each student recalls in each of the three categories. By ranking the results we may test whether recall ability differs systematically between categories, e.g. do people recall names of sporting personalities more easily than those of people prominent in local affairs? See Exercise 6.1.

Tasting tests

A number of people may be asked to taste several different varieties of raspberry and rank them in order of preference. A Friedman test detects any pattern in taste preference. See Exercise 6.4.

Quantal responses

Four doses of a drug are given to batches of rats, groups of four rats from the same litter forming a block. The only observation made is whether or not each treated rat is dead or alive after 24 hours have elapsed. Cochran's test is appropriate to test for different survival rates at different doses.

EXERCISES

6.1   At the beginning of a teaching period a list of 12 names is read out in random order to a class of 10 students. Four of the names are of prominent sporting personalities (group A), four of prominent national and international politicians (group B), and four of people prominent in local affairs (group C). At the end of the period students are asked to recall as many of the names as possible. The results are as follows:

Student	A	B	C	D	E	F	G	H	I	J
Group A	3	1	2	4	3	1	3	3	2	4
Group B	2	1	3	3	2	0	2	2	2	3
Group C	0	0	1	2	2	0	4	1	0	2

By ranking the above data within each block (student) use a Friedman test to determine whether there is a difference between recall rates for the three groups. In particular, is the recall rate for group B and/or group C significantly lower than that for group A?

The reader familiar with the standard parametric analysis of variance may like to carry out such an analysis on the original data. Do the conclusions agree with the Friedman test? If not, why not, and which analysis is to be preferred?

6.2   Use the rankings given in the literary discrimination example in Section 6.6 to determine whether the professor's claim to discriminate between the work of authors in a meaningful way is justified.

6.3   In traditional army fashion 34 men are paraded tallest on the right, shortest on the left, and numbered 1 (tallest) to 34 (shortest). This ranks the men by height. Each man is then asked by the sergeant major whether he smokes or drinks. The sergeant major notes the rank number of men in the various categories as follows:

Drinker and smoker	3 8 11 13 14 19 21 22 26 27 28 31 33
Smoker and non-drinker	2 12 25 32 34
Drinker and non-smoker	1 7 15 20 23 24 30
Non-smoker and non-drinker	4 5 6 9 10 16 17 18 29

Is there evidence of an association between height and drinking and smoking habits?

Would you reach a different conclusion if the analysis were carried out replacing ranks by van der Waerden scores? (In this preliminary analysis the factorial nature of the 'treatment' structure may be ignored.)

6.4   Five people are asked to rank four varieties of raspberry in order of taste preference. Ties may be awarded and specified as mid-ranks if a taster feels no distinction is possible between two or more varieties. Do the results indicate a consistent taste preference?

Taster	1	2	3	4	5
Variety					
Malling Enterprise	3	3	1	3	4
Malling Jewel	2	1.5	4	2	2
Glen Clova	1	1.5	2	1	2
Norfolk Giant	4	4	3	4	2

6.5   In uniform editions of works by each of three writers of detective fiction the number of sentences per page on randomly selected pages in each work are as follows:

C. E. Vulliamy	13 27 26 22 26
Ellery Queen	43 35 47 32 31 37
Helen McCloy	33 37 33 26 44 33 54

Use a Kruskal–Wallis test to determine whether it is reasonable to suppose there are differences in average sentence length between the authors. Which author appears to use the longest sentences? (Use of uniform editions ensures page and type sizes are the same for each author.)

If you are familiar with the parametric analysis of variance you may wish to carry out such an analysis on the above data. Does it lead to the same conclusions as the Kruskal–Wallis test? If not, why not, and which analysis is to be preferred?

6.6   Four share tipsters are each asked to predict on 10 randomly selected days whether the 100 share index will rise or fall on the following day. If they predict correctly this is scored as 1, if they predict incorrectly this is scored as 0. Do the scores recorded below indicate differences in the tipsters' abilities to predict accurately?

Day	1	2	3	4	5	6	7	8	9	10
Tipster 1	1	0	0	1	1	1	1	0	1	1
Tipster 2	1	1	1	1	0	1	1	0	0	0
Tipster 3	1	1	0	1	1	1	1	1	0	1
Tipster 4	1	1	0	0	0	1	1	1	0	1

6.7 If you are familiar with parametric analysis of variance, replace the data in Table 6.6 by ranks 1 to 24 (using mid-ranks for ties where appropriate) and carry out an ordinary randomized block analysis of variance to confirm the F-value quoted in Section 6.5.2.

6.8 Berry (1987) gives the following data for numbers of premature ventricular contractions per hour for 12 patients with cardiac arrhythmias when each is treated with 3 drugs.

Patient	1	2	3	4	5	6	7	8	9	10	11	12
Drug A	170	19	187	10	216	49	7	474	0.4	1.4	27	29
Drug B	7	1.4	205	0.3	0.2	33	37	9	0.6	63	145	0
Drug C	0	6	18	1	22	30	3	5	0	36	26	0

Use a Friedman test for differences in responses between drugs. In particular, is there evidence of a difference in response between drug A and drug B?

Note the obvious heterogeneity of variance between drugs. If you are familiar with the randomized block analysis of variance you may like to perform one on these data. Do you consider it is valid? Is the Friedman analysis to be preferred? Why?

6.9 Cohen (1983) gives data for numbers of births in Israel for each day in 1975. We give below data for numbers of births on each day in the 10th, 20th, 30th and 40th weeks of the year.

Day	Mon	Tues	Wed	Thur	Fri	Sat	Sun
Week							
10	108	106	100	85	85	92	96
20	82	99	89	125	74	85	100
30	96	101	108	103	108	96	110
40	124	106	111	115	99	96	111

Perform appropriate Friedman type analyses to determine whether the data indicate: (i) a difference in birth rate between days of the week that shows consistency over the four selected weeks, and (ii) any difference between rates in the 10th, 20th, 30th, 40th weeks.

6.10 Snee (1985) gives data on average liver weight per bird for chicks given three levels of growth promoter (none, low, high). Blocks correspond to different bird-houses. Use a Friedman test to see if there is evidence of an effect of promoter.

Block	1	2	3	4	5	6	7	8
None	3.93	3.78	3.88	3.93	3.84	3.75	3.98	3.84
Low dose	3.99	3.96	3.96	4.03	4.10	4.02	4.06	3.92
High dose	4.08	3.94	4.02	4.06	3.94	4.09	4.17	4.12

(Note that the dose levels are ordered so the Page test mentioned at the end of Section 6.5.1 would be appropriate.)

6.11 Lubischew (1962) gives measurements of maximum head width in units of 0.01 mm for three species of *Chaetocnema*. Part of his data are given below. Use a Kruskal–Wallis test to see if there is a species difference in head widths.

Species 1	53	50	52	50	49	47	54	51	52	57
Species 2	49	49	47	54	43	51	49	51	50	46	49
Species 3	58	51	51	45	53	49	51	50	51	

6.12 Biggins, Loynes and Walker (1987) considered various ways of combining examination marks where all candidates sat the same number of papers but different candidates selected different options from all those available. The data below give the marks awarded by 4 different methods of combining results for each of 12 candidates. Do the schemes give consistent ranking of the candidates? Is there any evidence that any one scheme treats some candidates strikingly differently to the way they are treated by other schemes so far as rank order is concerned? Is there any evidence of a systematic difference between marks awarded by the various schemes?

Candidate	1	2	3	4	5	6	7	8	9	10	11	12
A	54.3	30.7	36.0	55.7	36.7	52.0	54.3	46.3	40.7	43.7	46.0	48.3
B	60.6	35.1	34.1	55.1	38.0	47.8	51.5	44.8	39.8	43.2	44.9	47.6
C	59.5	33.7	34.3	55.8	37.0	49.0	51.6	45.6	40.3	43.7	45.5	48.2
D	61.6	35.7	34.0	55.1	38.3	46.9	51.3	44.8	39.7	43.2	44.8	47.5

7

Bivariate and multivariate data

7.1 CORRELATION IN BIVARIATE DATA

There are several nonparametric methods for correlation and regression. We consider first the bivariate case where methods are generally applied to a sample of bivariate observations (x_i, y_i), $i = 1, 2, \ldots, n$ that are at least ordinal. If they are measurements or counts we sometimes convert these to ranks and, in the case of correlation, base our statistics on these ranks. Depending on the precise problem, procedures for regression will sometimes use ranks and sometimes the original observations.

The oldest and perhaps most widely used measure of correlation is Pearson's product moment correlation coefficient

$$r = \frac{\sum_i (x_i y_i) - \left(\sum_i x_i\right)\left(\sum_i y_i\right)/n}{\sqrt{\left\{\left[\sum_i x_i^2 - \left(\sum_i x_i\right)^2/n\right]\left[\sum_i y_i^2 - \left(\sum_i y_i\right)^2/n\right]\right\}}} \qquad (7.1)$$

For a sample of n independent pairs of observations from a bivariate normal distribution many relevant tests and estimation procedures using r are readily available and many properties of r still hold if the distribution is not bivariate normal. For example, if the (x_i, y_i) when plotted lie almost on a straight line of positive slope (high values of x going with high values of y, low values of x with low values of y) then r will be close to $+1$. If the points lie exactly on such a straight line, $r = 1$. If all (x_i, y_i) lie close to a straight line of negative slope (high values of x go with low values of y and vice versa) then r has a value near -1; $r = -1$ exactly if the points are on such a line. For a complete scatter of points r is near zero. The converse is not necessarily true; for certain nonlinear patterns r may also be near zero. The inequality $-1 \leqslant r \leqslant 1$ always holds. Without the assumption of bivariate normality, hypothesis tests and estimation procedures for r depend on the distribution of x and y and are thus not distribution-free.

Two well-known nonparametric measures of correlation based on ranks may be applied either to data which consists *ab initio* of ranks or to ranks derived from continuous data or counts. The first is **Spearman's rho** (denoted by the Greek letter ρ). This is just Pearson's product moment coefficient calculated for the ranks (with ties represented by mid-ranks). A second commonly used measure is **Kendall's tau** (denoted by the Greek letter τ), and there is no obvious parametric analogue. Both ρ and τ share with r the properties that they lie between $+1$ and -1, taking the value $+1$ when the ranks of x and y agree completely, -1 when they are precisely opposite (e.g. in 10 observations rank 1 for x is paired with rank 10 for y, rank 2 for x with rank 9 for y, and so on). If the ranks show no obvious relationship, values of ρ or τ near zero may be expected.

7.1.1 Spearman's rank correlation coefficient

If we replace the (x_i, y_i) by their ranks we may calculate ρ for these ranks using (7.1). The simplified formula (7.3) uses properties of ranks. Writing (r_i, s_i) for the ranks of (x_i, y_i), then, whether or not there are ties involving mid-ranks, a formula for ρ is

$$\rho = \frac{\sum_i r_i s_i - C}{\sqrt{\left\{\left[\sum_i r_i^2 - C\right]\left[\sum_i s_i^2 - C\right]\right\}}} \tag{7.2}$$

where $C = \frac{1}{4}n(n+1)^2$. If there are no ties (7.2) reduces to

$$\rho = 1 - \frac{6T}{n(n^2-1)} \tag{7.3}$$

where $T = \sum_i [r_i - s_i]^2$, i.e. T is the sum of the squares of the differences between the ranks of x and y for each observation. For only a few ties there will be little difference between ρ calculated by (7.3) or (7.2), so it is common practice to use (7.3) as a good approximation in this case although the general applicability of (7.2) makes it preferable for a computer routine for calculating ρ. Clearly, with perfect matching of ranks T is zero, so $\rho = 1$. If there is an exact reversal of ranks, tedious but elementary algebra shows that $\rho = -1$. If there is no correlation between ranks it can be shown that the expected value of T is $n(n^2-1)/6$, so that ρ then has expectation zero. If our observations are a random sample from a bivariate distribution with X, Y independent we expect near zero rank correlation. Table A9 gives critical values for testing $H_0: \rho = 0$ against one- or two-sided alternatives. Since ρ in (7.3) is a function of T only for fixed n, T itself may be used as a test statistic and has been tabulated for small n. We prefer to use ρ as it gives a more easily appreciated indication of the level of correlation. For $n > 20$ the approximation given in Table A9 is usually adequate.

A common use for ρ is to decide if there is broad agreement between ranks assigned by two different assessors. Do two examiners agree in their ranking of candidates? Do two judges agree in their placings in a beauty contest? Do job applicants' ranks for manual skill on the basis of some psychological test show any relation to their ranking in a further test for mental skills?

The coefficient ρ was proposed first by Spearman (1904) and is also useful in tests for trend; it is often more powerful than the Cox–Stuart test described in Section 2.1.6.

Example 7.1

The problem. Life expectancy has shown a general increase during the nineteenth and twentieth centuries as standards of hygiene and health care have improved. The extra life expectation varies from nation to nation, community to community and even family to family. Table 7.1 gives the year of death and age at death for 13 males of a fourth clan, the McDeltas, buried in the Badenscallie burial ground. Is there any indication that life expectancy is increasing for members of this clan?

Table 7.1 Year of death and ages of 13 McDeltas

Year	1827	1884	1895	1908	1914	1918	1924	1928
Age	13	83	34	1	11	16	68	13

Year	1936	1941	1964	1965	1977
Age	77	74	87	65	83

Formulation and assumptions. If there is an increase in life expectancy we expect those dying later to tend to be older than those dying earlier. If x is the year of death and y the age, then if we rank the x and y, an increase in life expectancy should be indicated by a positive ρ; if this is sufficiently large we reject $H_0: \rho = 0$ in favour of the alternative $H_1: \rho > 0$. A one-sided test is appropriate if we accept that there is little likelihood of life expectancy decreasing in the light of national trends.

Procedure. The years already are ordered. We replace these by ranks and also replace each age of death by the appropriate rank, giving the rank pairings in Table 7.2. We add an additional line that gives the difference in ranks for each pair.

Table 7.2 Ranks for McDelta death data

Year rank (r)	1	2	3	4	5	6	7
Age rank (s)	3.5	11.5	6	1	2	5	8
$r - s$	− 2.5	− 9.5	− 3	3	3	1	− 1

Year rank (r)	8	9	10	11	12	13
Age rank (s)	3.5	10	9	13	7	11.5
$r - s$	4.5	− 1	1	− 2	5	1.5

To calculate T we square and add the differences in the final row of Table 7.2, giving $T = 179$; and from (7.3), $\rho = 0.508$. When $n = 13$, Table A9 gives 0.484 for the critical value for significance in a one-tail test at the 5% level.

Conclusion. Since $0.508 > 0.484$ we reject the hypothesis that there is no increase in life expectancy with time.

Comments. 1. In Exercise 7.1 we examine whether the Cox–Stuart trend test would reject the null hypothesis.

2. One might anticipate that the larger samples for the McAlphas, McBetas and McGammas discussed in earlier chapters would also give significant results. It turns out that for each of these groups although the calculated ρ is positive, it just fails to reach significance at the 5% level in each case. We may have been somewhat lucky to get a significant result with our small sample of McDeltas. We already drew attention in Example 3.4 to the bimodal form of death statistics due to a tendency for death to occur in infancy or else after survival to fairly mature years. In the case of the McDeltas, the several deaths before maturity tended to occur fairly early in the time sequence. The result would clearly have been negated had there been a death in infancy in the 1980s for that family.

3. The fact that ρ for each of the four clans was positive (though not significant) suggests a trend towards increasing life expectation. Indeed, if available data are combined for all clans (though there are dangers in combining data if they are in fact samples from different populations) we should enhance our prospect of a significant result. For the combined sample of 115 (death dates were not available for 2 individuals) Spearman's $\rho = 0.2568$, significant at the 1% level in a one-tail test.

4. If, for the McDeltas, we use (7.2) to allow for ties we find $\rho = 0.5069$ and our conclusion is unaltered.

5. While the test result formally supports the hypothesis of increasing life expectancy we must bear in mind that not all members of a clan will be buried near their place of birth; those that are may not be typical of the whole clan. Some will be lost by emigration and this may not have taken place at a steady rate throughout the period. The famous (or infamous) Highland clearances had an impact on local death statistics for several generations. There may also be an age, or change with time, element that determines preferences for cremation as opposed to burial. Such points must be kept in mind when deciding the 'real world' implications of a statistically significant result.

7.1.2 Kendall's rank correlation coefficient

M. G. Kendall (1938) proposed a coefficient with similar properties to Spearman's ρ, but a different logical basis. Noether (1981) suggests it is easier to interpret than Spearman's ρ. Kendall noted that if there is a tendency for high x values to be associated with high y values and we have observations (x_i, y_i) and (x_j, y_j), then if $x_i > x_j$ it is very probable that $y_i > y_j$. Pairs of observations having this property are described as **concordant**. Thus the observations (2.3, 7.1) and (3.5, 8.2) are concordant since 3.5 is greater than 2.3 and 8.2 is greater than 7.1. Concordance implies the differences $x_j - x_i$ and $y_j - y_i$ both have the same sign, or equivalently that $(y_j - y_i)/(x_j - x_i)$ is positive. If the signs of the numerator and denominator differences are opposite, (i.e. the quotient is negative) the pair is said to be **discordant**. Thus (3, 4) and (6.5, 1.5) are discordant since 6.5 is greater than 3 but 1.5 is less than 4. If $x_i = x_j$ or $y_i = y_j$ (or both) the corresponding pair are neither concordant nor discordant and we call this a **tie**. In all, for n observations there are $\frac{1}{2}n(n-1)$ possible pairings. If there are n_c concordant pairs, n_d discordant pairs, and t ties, $n_c + n_d + t = \frac{1}{2}n(n-1)$. Kendall's tau is computed as

$$\tau = \frac{n_c - n_d}{\frac{1}{2}n(n-1)} \tag{7.4}$$

If all pairs are concordant clearly $n_c = \frac{1}{2}n(n-1)$ and $\rho = 1$, while if all are discordant $n_d = \frac{1}{2}n(n-1)$ and $\rho = -1$. If there is independence we expect n_c, n_d to be roughly equal and ρ will be close to zero. Note that for both Spearman's ρ and Kendall's τ, values of ± 1 imply linearity in the ranks (but not necessarily in the original variables on which ranks may be based). Values of ± 1 for the Pearson coefficient, r, on the other hand, imply linearity for the original variables.

Kendall's τ may be calculated using either pairwise ranks or the original data. It is best to arrange data in ascending order for one of the variables (preferably x), noting any ties in the x values. One then need only count for each x_i the numbers of y that exceed y_i in pairs occurring later in the hierarchy, for these will be concordant. Those with lower y values will be discordant and those with equal y values will be ties (as will pairs with equal x values). The count is made systematically for all possible pairs in the way described in Example 7.2. Tables exist for critical values of τ itself, e.g. Neave (1981, p. 40), but because they are more useful for some other applications, tables giving critical values not for τ itself but for $n_c - n_d$ are also common. Table A10 is one such table; although use of this table does not necessitate evaluating τ, it is advisable to do so as it gives a 'feel' for the closeness of association. Some tables (e.g. Conover, 1980, Table A12) give quantiles that must be exceeded for significance. This is quite satisfactory if one notes the distinction between critical values and quantiles when nominal and actual significance levels do not coincide.

Example 7.2

The problem. Use the data in Table 7.1 to compute Kendall's τ. Compare your result with that for Spearman's ρ in Example 7.1.

Formulation and assumptions. The data in Table 7.1 are already ordered by the x variate (year of death) so we need only compare y's in much the way we proceed in the Mann–Whitney formulation of the Wilcoxon–Mann–Whitney test.

There are no ties among the x (years). The first y entry is 13. In the following entries this is exceeded in 9 cases (concordances since all x's exceed the first when we order by x). There is 1 tie and 2 lower y values representing discordances. To tally, we enter the numbers of concordances and discordances in the first line of Table 7.3.

Next we start with $y = 83$ corresponding to $x = 1884$ and examine the succeeding y. We find 9 discordances (smaller y), 1 concordance, and 1 tie in this case. We enter the numbers of concordances and discordances in the second line of Table 7.3. The third y value of 34 is followed by 6 concordances and 3 discordances. Proceeding in this way we finally come to $y = 65$ in 1965 which is followed only by $y = 83$, representing a concordance, so the entries 1 and 0 are made in Table 7.3 before we add to get n_c and n_d.

Table 7.3 Concordances and discordances

Concordances	Discordances
9	2
1	9
6	4
9	0
8	0
6	1
4	2
5	0
2	2
2	1
0	2
1	0
$n_c = 53$	$n_d = 23$

From Table 7.3 we find $n_c - n_d = 30$ and hence from (7.4) $\tau = 30/78 = 0.3846$. From Table A10 we find that $n_c - n_d$ must exceed 28 for significance at the 5% level in a one-tail test.

Conclusion. Since we observe $n_c - n_d = 30$, we reject at the 5% level the null hypothesis that life expectancy is not increasing.

Comments. 1. The value of Kendall's τ is lower than that of Spearman's ρ for these data, yet the significance level is almost identical (in neither case would we reject H_0 in a two-tail test at the 5% level). For most data sets one finds $|\rho| > |\tau|$, but the conclusion regarding significance is the same in either case. Which coefficient one uses is often governed by ease in calculation or availability of an appropriate computer program.

2. A little care is needed in counting concordances if there are ties in both x and y. For example, for the data set:

$$x \quad 2 \quad 5 \quad 5 \quad 7 \quad 9$$
$$y \quad 5 \quad 1 \quad 3 \quad 8 \quad 1$$

the first pair (2, 5) gives 1 concordance (7, 8) and 3 discordances; the second pair (5, 1) gives 1 concordance (7, 8) and no discordances because of the tied values at (5, 3) and again at (9, 1); the next pair (5, 3) gives one concordance (7, 8) and one discordance (9, 1).

A probabilistic interpretation of Kendall's τ as a measure of the probability that randomly selected pairs are placed in correct order is given by Kerridge (1975).

7.1.3 A graphical method

There is an easy graphical procedure for calculating Kendall's τ if there are no ties.

Example 7.3

The problem. Ten students are ranked by their marks in French and German examinations, the student with the highest mark in each being ranked 1. The ranks are:

Student	A	B	C	D	E	F	G	H	I	J
French rank	1	2	3	4	5	6	7	8	9	10
German rank	1	3	2	7	5	4	8	6	9	10

Use a graphical method to calculate Kendall's tau.

Formulation and assumptions. The students must (as above) be ranked in order for x (here the French mark). The graphical procedure below enables us to determine immediately the number of discordances; since there are no ties, the number of concordances is given by $n_c = \frac{1}{2}n(n-1) - n_d$, so we may easily obtain $n_c - n_d$.

Procedure. The ranks are written down in parallel rows in the same order as they occur above (see Figure 7.1).

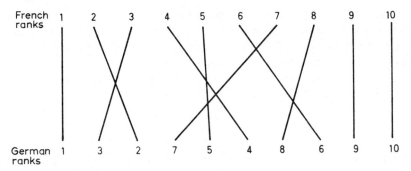

Figure 7.1 A graphical method for calculating τ

Corresponding ranks in the top and bottom row are joined. The number of intersections of these joins gives n_d. Care should be taken in making the joins to avoid coincident intersections; this is easily achieved by slight displacement of lines if necessary.

From Figure 7.1 we see that there are six intersections, thus $n_d = 6$. Since $n = 10$ it follows that $n_c = \frac{1}{2} \times 10 \times 9 - 6 = 39$.

Conclusion. Using (7.4) we easily find $\tau = (39 - 6)/45 = 0.733$.

Comments. 1. The reader should, by a direct count of the number of discordances, verify that the result obtained by the graphical method is correct.

2. Since $n_c = \frac{1}{2}n(n-1) - n_d$ when there are no ties, it is only necessary to count the number of concordances or discordances in the 'no ties' case when using either the method in this example, or that in Example 7.2.

An interesting application of the graphical procedure to actual and predicted association football results for the English and Scottish leagues is given by Hill (1974).

7.2 NONPARAMETRIC BIVARIATE LINEAR REGRESSION

The classic method for fitting a straight line to bivariate data is the method of least squares. This has optimal properties subject to certain independence and normality assumptions; procedures for hypothesis testing and estimation under these assumptions are well known.

Given a set of n bivariate observations (x_i, y_i) there are often situations where the assumptions needed to validate least squares methods for hypothesis tests or estimation do not hold. This may result in invalid or misleading inferences.

In Section 5.1.2, Example 5.3, we showed that the t-test may be insensitive with samples that are clearly not from a normal distribution. In the regression situation similar problems arise. While the method of least squares may be relatively insensitive to some departures from basic assumptions, it can be grossly affected by others. In the past two decades many diagnostic tests have been developed to detect observations that may cause difficulties with standard regression methods, or observations that are very influential in the sense that relatively small changes in them may have a marked effect on estimates. At the same time robust methods have been developed that are little affected by aberrant observations but behave almost as well as least squares when that method is appropriate. We say more about one important class of robust methods in Chapter 9 – the so-called m-estimators – but in this chapter we explore a procedure closely related to Kendall's τ. Classical least squares methods are related to Pearson's product moment correlation coefficient. A key difference between correlation and regression is that correlation is concerned solely with **strength** of a linear (or linear in ranks) relationship, whereas regression problems deal with the **precise form** of a linear (or some other) relationship. We first say something about the nature of regression and highlight by an example one difficulty with least squares estimation when basic assumptions are violated.

7.2.1 Least squares regression

Most experimenters meet bivariate least squares regression as the method that gives, in some sense, the line or curve that best fits their data. The assumption usually made is that we have n paired observations (x_i, y_i), $i = 1, 2, \ldots, n$. The x_i may be values of some mathematical variable, e.g. dates, or times after application of a treatment, the dates or times being either at the experimenter's choice or randomly selected. It is further assumed these are measured without error, or at least that any measurement error is negligible. The y_i in a regression problem are always random variables. What an experimenter often assumes, when he says that data lie almost on a straight line, is that if it were not for some nuisance variation (in a biological situation this might be genetic variability; in a manufacturing process machine variability or varying quality of raw material; for a psychologist the vagaries of the human mind) the points would lie exactly on a straight line. The experimenter does not usually know what that straight line is. A plot of the observations may produce a graph that looks like Figure 7.2.

The least squares fit will be valid if we assume that for any given x, say $x = x_i$, our y_i is an observed value of a random variable Y_i which has a normal distribution with mean conditional on x_i of the form $E(Y_i|x = x_i) = \alpha + \beta x_i$ and (usually unknown) variance σ^2, which is independent of x_i. Here the notation $E(Y_i|x = x_i)$ means the expectation of Y_i **conditional upon** (or when) x

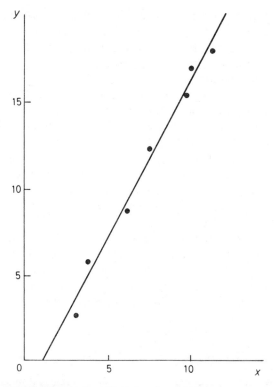

Figure 7.2 Scatter diagram and least squares regression line

takes the value x_i; here this mean equals $\alpha + \beta x_i$. The regression problem is to estimate the unknown constants α, β (and sometimes σ^2). Another way of stating the regression problem is to say that for a given (x_i, y_i) we may write

$$y_i = \alpha + \beta x_i + \varepsilon_i \tag{7.5}$$

where ε_i is an unobservable random variable that has a normal distribution with mean 0 and variance σ^2.

The statistician also meets the linear least squares regression problem as that of estimating the conditional mean of Y for given X when given observations from a bivariate normal distribution where X and Y are correlated. This estimation problem has the same solution as that for fitting a straight line in the experimental situation described above, but the statistician may seek a solution in cases where the data do not necessarily lie close to a straight line; however, the locus of the conditional mean is still that straight line. We shall not pursue this problem here, but return to the more practical research version described above. Least squares gives estimates a, b of α, β that minimize $\sum_i (y_i - a - bx_i)^2$, i.e. the sum of squares of deviations of the observed points from the fitted line when these deviations are measured perpendicular to the x-axis (or parallel to the y-axis). For data like those in Figure 7.2 the fit so obtained is good in any reasonable sense of the word and is shown on that figure. However, there are cases where a least squares fit may give concern.

Example 7.4

The problem. Use the method of least squares to fit a straight line to the following points:

$$
\begin{array}{llllllll}
x & 0 & 1 & 2 & 3 & 4 & 5 & 6 \\
y & 2.5 & 3.1 & 3.4 & 4.0 & 4.6 & 5.1 & 11.1
\end{array}
$$

Formulation and assumptions. We show in Section A6 that the estimators a, b are

$$
b = \left[\sum_i (x_i y_i) - \left(\sum_i x_i \right)\left(\sum_i y_i \right) \Big/ n \right] \Big/ \left[\sum_i (x_i^2) - \left(\sum_i x_i \right)^2 \Big/ n \right]
$$

and

$$
a = \bar{y} - b\bar{x}
$$

where $\bar{x} = (\sum_i x_i)/n$, $y = (\sum_i y_i)/n$.

Procedure. We find $n = 7$, $\sum_i x_i = 21$, $\sum_i y_i = 33.8$, $\sum_i x_i^2 = 91$, $\sum_i (x_i y_i) = 132.4$, whence $b = 1.11$ and $a = 1.50$.

Conclusion. The least squares regression line is

$$
y = 1.50 + 1.11x
$$

Comments. The points are plotted in Figure 7.3 along with the least squares regression line.

The fit is disturbing. If we substitute the given values of x in our calculated equation we get an estimate of the mean value of y conditional on each of these x values. The experimenter using this method regards these as the 'best' estimates of the idealized y if there were no random variation. We often denote these estimates by $\hat{y}$. If we calculate for each data point $y_i - \hat{y}_i$ we obtain the **residuals** at the data points. For our data the residuals in the order of the data are 1.00, 0.49, -0.32, -0.83, -1.34, -1.95, 2.94. The magnitude of the residuals is given by the vertical distance of each point to the line; the sign being negative if the observation lies below the line, positive if it is above (see Figure 7.3).

If our fitted line were the true unknown line the residuals would correspond to the unknown ε_i in equation (7.5). Therefore, there is a sense in which we might regard the residuals $y_i - \hat{y}_i$ as estimates of the realized values of a random variable ε. If the 'errors' are independent of x – an assumption in least

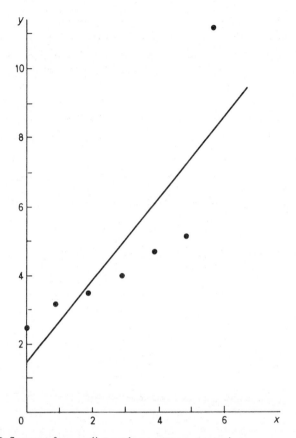

Figure 7.3 Influence of an outlier on least squares regression.

squares – it is reasonable to expect a more or less random pattern of positive and negative residuals. That is clearly the situation in Figure 7.2, but if we look at the above residuals in order of increasing x we see the first two are positive and decreasing in magnitude, then they change sign and progressively increase in magnitude until we come to that corresponding to $x = 6$ when we get a swing to a large positive residual. In Figure 7.4 we plot the residuals against the x_i. They have a very definite relationship to the x_i.

Clearly here we have an anomaly as the 'estimated errors' or residuals are not independent of the x_i. This is because 6 of our 7 points lie close to a straight line, but one is well away from the line suggested by these other points. There are two possible conclusions. Either a straight line with the given homogeneous error assumptions is not adequate for our data or there is something peculiar about our last observation. It is indeed about double the value we would expect on the basis of the other points. This sort of 'data mishap' is not infrequent. If the y values are the average of two measurements, did somebody

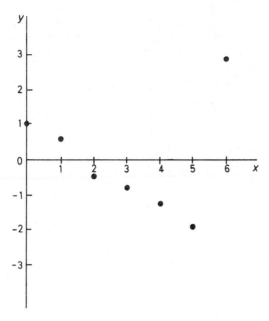

Figure 7.4 Plot of residuals against x for data in Example 7.4.

forget to divide by two in this last case? The data itself cannot tell us, but a data feature like this suggests we should make enquiries about it.

It is problems like this that have stimulated recent interest in so-called regression diagnostics; in particular, developing methods to detect suspect observations or influential points. For a full discussion see Cook and Weisberg (1982) or Atkinson (1986). The problem introduced here becomes more important in multiple regression (where there are several independent or regressor variables); simple graphical techniques are then not easily available and elementary studies of residuals may no longer suffice. It is for these situations that diagnostic techniques are especially important.

An alternative approach to detect possible difficulties in bivariate (and multiple) regression is to fit by least squares and also by a robust method which weighs down the influence of any suspect point. If the two methods agree there is unlikely to be a problem. If there is a large discrepancy between results using two methods for the same data, one must explore how the difference arises. Good robust methods give an answer like least squares when that is appropriate, so it is sometimes argued that we should always use robust methods and simply accept the answer. One reason for not accepting this argument is that fitting a straight line may be the wrong thing to do for some data; even a robust method may not reveal this without further diagnostic tests. Sometimes we need more data to resolve an apparent anomaly. If we had some additional points to those in Example 7.4 with x values between 5 and 6 we might end up with a plot like Figure 7.5. In this case it is clear some nonlinear curve would fit our data. Without such data, and when no reason could be given by an experimenter for the odd behaviour of the point at $x = 6$, then it is sensible to collect more data if feasible. If it is only possible to take observations at integral x values every effort should be made to get a repeat measurement at $x = 6$, and if possible additional ones at $x = 7$ or $x = 8$ to evaluate the continuing trend.

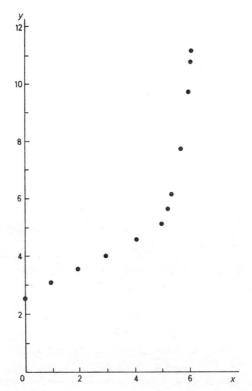

Figure 7.5 A relationship that is clearly nonlinear.

7.2.2 Theil's regression method

Theil (1950) proposed estimating the slope of a regression line as the median of the slopes of all lines joining pairs of points with different x values. For a pair (x_i, y_i) and (x_j, y_j) the relevant slope is

$$b_{ij} = \frac{y_j - y_i}{x_j - x_i} \tag{7.6}$$

Suppose all x_i are different; it is convenient to order observations in ascending order of x. Clearly $b_{ij} = b_{ji}$ for all i and j so for n observations there are $\frac{1}{2}n(n-1)$ algebraically distinct b_{ij} and it is convenient to write them in an upper triangular matrix:

$$
\begin{matrix}
b_{12} & b_{13} & b_{14} & . & . & . & . & . & b_{1n} \\
 & b_{23} & b_{24} & . & . & . & . & . & b_{2n} \\
 & & b_{34} & . & . & . & . & . & b_{3n} \\
 & & & . & . & . & . & . & . \\
 & & & & & & & & b_{n-1,n}
\end{matrix}
$$

The calculated numerical matrix has pattern and even with no suitable computer program the median is usually easy to detect for moderate n without further ordering. The process is somewhat reminiscent of that for obtaining the Hodges–Lehmann estimator in the Wilcoxon signed rank procedure. In Section 7.2.4 we show that it is related to Kendall's τ.

If we denote the median estimator of β by $\tilde{b}$, Theil suggested estimating α by $\tilde{a}$, the median of all $a_i = y_i - \tilde{b}x_i$; or alternatively we might choose $\tilde{a}' = \text{med}(y_i) - \tilde{b}\,\text{med}(x_i)$, where $\text{med}(x_i)$ is the median of all observations. If we use the latter our fitted line passes through the median of all observations, whereas the least squares line passes through the mean.

Example 7.5

The problem. Estimate the straight line regression of y on x for the data in Example 7.4 using Theil's method.

Formulation and assumptions. We calculate the b_{ij} and their median. We then calculate the median of all a_i as defined above.

Procedure. The data are:

x	0	1	2	3	4	5	6
y	2.5	3.1	3.4	4.0	4.6	5.1	11.1

We first calculate

$$b_{12} = (y_2 - y_i)/(x_2 - x_1) = (3.1 - 2.5)/(1 - 0) = 0.6$$

and proceed in this way, writing the results in matrix form:

0.600	0.450	0.300	0.525	0.520	1.433
	0.300	0.450	0.500	0.500	1.600
		0.600	0.600	0.567	1.900
			0.600	0.550	2.367
				0.500	3.250
					6.000

Clearly there are 21 b_{ij} in this matrix so the median is the 11th largest. Inspection shows $\tilde{b} = 0.567$. To estimate $\tilde{a}$ we calculate $a_1 = 2.5 - 0.567 \times 0 = 2.5$, and similarly $a_2 = 2.533$, $a_3 = 2.266$, $a_4 = 2.299$, $a_5 = 2.332$, $a_6 = 2.265$, $a_7 = 7.698$. The median of the a_i is $\tilde{a} = 2.332$.

Conclusion. The fitted line is $y = 2.332 + 0.567x$. Figure 7.6 shows this line together with the data points.

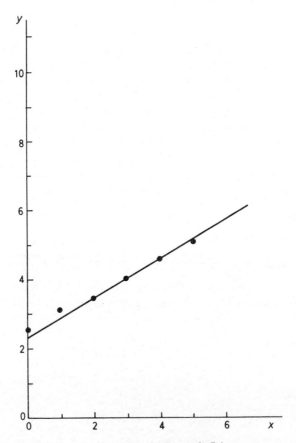

Figure 7.6 Theil regression fitted to data in Example 7.4.

Comments. 1. It is clear from Figure 7.6 that the point (6, 11.1) has a much reduced influence in determining the line of best fit. This is because the other pairings all give reasonably consistent values of b_{ij} ranging between 0.3 and 0.6, only the six b_{ij} involving the suspected 'rogue' point give wildly different values (ranging from 1.433 to 6.00). Their presence lifts the median slightly. Had we not had this wild observation the median for all other points is 0.520.

2. If we estimate α by $\tilde{a}' = \text{med}(y_i) - \tilde{b}\,\text{med}(x_i)$ we find $\tilde{a}' = 2.299$, reasonably close to the value given by the Theil procedure.

3. When there are no rogue observations, intuitively one feels that more weight should be given to b_{ij} for pairs of points fairly distant from each other than one would give to b_{ij} based on points close together. This is an argument for taking some sort of weighted median. Jaeckel (1972) proposed weighting each b_{ij} with a weight

$$w_{ij} = (x_j - x_i) \bigg/ \sum_{i \neq j} (x_j - x_i)$$

and recommended taking the median b_{ij} with these assigned weights. The procedure has certain optimal properties when there are no outliers but this and other weighting schemes may lead to slightly less robust estimators when there are outliers. Alternative weighting schemes have been discussed by Scholz (1978), Sievers (1978) and Kildea (1978). In passing it is interesting to note that the least squares estimator is equivalent to the weighted mean of the b_{ij} with weights proportional to $(x_j - x_i)^2$.

4. For large n computing problems become formidable with a pocket calculator, so unless a computer program for the Theil procedure is available there is a case for seeking some short-cut method. One is given in Section 7.2.3.

The fact that the data in Examples 7.4 and 7.5 lead to very different fits using least squares and Theil's method does not solve our problem as to which is the appropriate fit. Example 7.4 suggests that the least squares fit may not be appropriate in that the data are unlikely to be near a straight line with normal and identically distributed errors; but leaves unresolved the question whether the point (6, 11.1) is seriously in error, or the relationship is not truly linear. More data might answer that question; so might a check on the data (e.g. did someone forget to divide by 2?). If we decide there is reason to ignore the last point we might use a least squares fit on the remaining points or be content to use the result from Theil's procedure. Similar results could be obtained by other robust methods using principles outlined (but not described in detail) in Section 9.1.4.

In simulation studies involving outliers that effectively amounted to replacing normally distributed errors with errors from long-tail distributions, Hussain and Sprent (1983) found that Theil's method was almost as efficient as least squares when the normality assumptions were valid and that it showed a marked improvement in efficiency with a long-tail error distribution, especially with sample sizes less than 30. There was an even more marked improvement in the latter case in the estimates of α, although this is commonly of less interest than β.

Hussain and Sprent also found that estimators based on weighted medians performed on the whole no better, and sometimes less well, than Theil estimators in the presence of outliers.

7.2.3 The abbreviated Theil method

If suitable computational facilities are not available for the complete procedure, Theil suggested the following abbreviated procedure for n distinct x values. If n is even, we take $\tilde{b} = \mathrm{med}(b_{i,i+(1/2)n})$, $i = 1, 2, \ldots, \frac{1}{2}n$; and if n is odd we take $\tilde{b} = \mathrm{med}(b_{i,i+(1/2)(n+1)})$, $1 = 1, 2, \ldots, \frac{1}{2}(n-1)$. The procedure is not generally recommended unless $n \geqslant 12$. If $n = 14$, say, we compute only $b_{1,8}$, $b_{2,9}, b_{3,10}, b_{4,11}, b_{5,12}, b_{6,13}, b_{7,14}$. For odd n the procedure is very similar and all observations except that corresponding to $x_{(1/2)(n+1)}$, where the x are ordered, are used.

The method is often called the **abbreviated Theil method**. We illustrate it using the data in Example 7.5 even though the number of points fall below our suggested minimum for the procedure to be reasonably efficient. It is usual to estimate α as for the full Theil method as this involves only n sets of easy computations.

Example 7.6

The problem. Use the data in Examples 7.4 and 7.5 to calculate the straight-line regression by the abbreviated Theil method.

Formulation and assumptions. For 7 points we calculate only b_{15}, b_{26}, b_{37}, and obtain their median.

Procedure. From the matrix in Example 7.5 we find $b_{15} = 0.525$, $b_{26} = 0.500$, $b_{27} = 1.900$. Their median is 0.525. This differs from the value of $\tilde{b}$ obtained in Example 7.5; the estimate of α will also differ. It is left as an exercise to show that now $\tilde{a} = 2.5$.

Conclusion. The estimated regression line is

$$y = 2.5 + 0.525x$$

Comments. Although the line is close to that obtained by the full Theil method, the fact that the estimate is the median of only three b_{ij} could, with some data sets, cause difficulty. We shall see in Section 7.2.4 also that it is not possible to obtain a sensible confidence interval for β with so few b_{ij}.

7.2.4 Hypothesis tests and confidence intervals using Theil's method

Clearly a good estimator b of β should be such that the residuals associated with each observation, denoted by e_i where $e_i = y_i - a - bx_i$, should be equally likely to be positive or negative. This implies an assumption that the e_i are randomly distributed with a zero median and independent of the x_i. Now

$$b_{ij} = \frac{y_j - y_i}{x_j - x_i} = \frac{a + bx_j + e_j - (a + bx_i + e_i)}{x_j - x_i} = b + \frac{e_j - e_i}{x_j - x_i} \qquad (7.7)$$

Equation (7.7) implies any b_{ij} will be greater than b if (x_i, e_i) and (x_j, e_j) are concordant and that b_{ij} will be less than b if these are discordant in the sense used in Kendall's τ, Section 7.1.2. Choice of the med$\{b_{ij}\}$ as estimator of b ensures half the pairs are concordant and half discordant. Any choice of b for which we accept Kendall's τ consistent with zero correlation between the observed x and the corresponding residuals, e, is acceptable in the sense that it is consistent with zero correlation between x and the residuals. In other words, we accept any b which does not give a number of discordants (or concordants) that indicate Kendall's τ differs from zero, i.e. we do not want the number of discordants (or concordants) too small or too large. Since $n_c + n_d = N$ equals the total number of b_{ij} generated from n observations with distinct x_i, we reject $\tau = 0$ in a two-tail test at the 5% level if $|n_c - n_d|$ is greater than or equal to the critical value given in Table A10.

This enables us to form, say, a nominal 95% confidence interval for β. Without loss of generality we assume $n_c > n_d$. Then if c is the critical value for testing at the 5% significance level in a two-tail test we reject any hypothetical value of β for which $n_c - n_d \geqslant c$. But since $n_c + n_d = \frac{1}{2}n(n-1) = N$, say, it follows that we reject all estimators for which $n_d \leqslant \frac{1}{2}(N-c)$. Thus the critical value of n_d is $r = \frac{1}{2}(N-c)$. If this is non-integral we round down. It follows that the lower limit to our confidence interval is given by the $r + 1$ smallest b_{ij}. By symmetry the upper limit is obtained by rejecting the r largest.

Example 7.7

The problem. For the data of Example 7.5 obtain a 95% confidence interval for β.

Formulation and assumptions. We use the method based on Kendall's τ described above.

Procedure. Entering Table A10 for $n = 7$ we find for a two-tail test at the 5% level the critical value for $|n_c - n_d|$ is 15. Now N (the number of b_{ij}) is 21, thus $r = \frac{1}{2}(21 - 15) = 3$, so we reject the 3 largest and 3 smallest b_{ij}; the next smallest and largest give confidence limits, so the 95% confidence interval for β is the interval $\{(b_{ij}(4), b_{ij}(18)\}$ where $b_{ij}(4)$, $b_{ij}(18)$ are the 4th and 18th ordered b_{ij} computed in Example 7.5. From the values found in that example we easily see $b_{ij}(4) = 0.450$ and $b_{ij}(18) = 1.900$.

Conclusion. A 95% confidence interval for β is (0.450, 1.900).

Comments. 1. The procedure is reminiscent of that for confidence intervals for the Wilcoxon signed ranks or Wilcoxon–Mann–Whitney situations. The point estimate med(b_{ij}) is the Hodges–Lehmann estimator.

2. Note that unlike the conventional confidence intervals for β using least

squares under an assumption of normality this interval is not symmetric about the point estimator $\text{med}(b_{ij})$. Were it not for the observation at $x = 6$ being 'out of line' with the other observations a more symmetric interval would have been obtained. Lack of symmetry draws attention to the outlier. However, in some situations there may be one or more outliers calling for a robust method, but confidence intervals may still be reasonably symmetric.

3. The logic behind least squares and Theil's method is basically different. The least squares estimator may sometimes (but not in this case; see Example 7.4) be outside the Theil confidence interval.

Sen (1968) extends the above arguments for confidence intervals to situations where we have tied x or e values. Conover (1980, Section 5.5) discusses hypothesis tests for least squares estimates with assumptions that do not include normality and these depend on Spearman's ρ rather than Kendall's τ.

For the abbreviated Theil method confidence intervals are easily calculated because the b_{ij} involved are all calculated for different point pairs i and j and are therefore mutually independent. Thus for an acceptable estimator each relevant $(e_j - e_i)/(x_j - x_i)$ has probability $\frac{1}{2}$ of being positive and probability $\frac{1}{2}$ of being negative. If we use N' independent b_{ij} in our modified Theil procedure, we set confidence limits on β in exactly the way we set confidence intervals for the median with a sample of N' independent observations when using the sign test (see Section 2.1.2).

Example 7.8

The problem. For 18 points the abbreviated Theil estimators are $b_{1,10} = 1.96$, $b_{2,11} = 1.89$, $b_{3,12} = 2.18$, $b_{4,13} = 2.04$, $b_{5,14} = 1.87$, $b_{6,15} = 2.01$, $b_{7,16} = 2.02$, $b_{8,17} = 2.07$, $b_{9,18} = 2.57$. Determine the 95% confidence interval for β.

Formulation and assumptions. We rearrange the 9 b_{ij} in ascending order and use Table A1 to determine the confidence interval treating the ordered b_{ij} as a sample of 9 for which we wish to determine a confidence interval for the median as in the sign test.

Procedure. The 9 ordered b_{ij} are 1.87, 1.89, 1.96, 2.01, 2.02, 2.04, 2.07, 2.18, 2.57. From Table A1 with $N = 9$ we find we reject a median value that results in 1 or fewer or 8 or more minus signs when testing at a nominal 5% level (actual level 3.92%). If our estimate satisfies $b_0 > 1.89$ and $b_0 < 2.18$ clearly we would accept the value.

Conclusion. The 95% confidence interval for β is (1.89, 2.18).

Comment. If we used the full Theil estimation procedure we could reasonably expect the confidence interval based on Kendall's τ to be shorter, as we are using a greater amount of relevant information.

7.2.5 A test for parallelism

If we use the abbreviated Theil method to fit lines to two data sets we get a simple test for parallelism or concurrence. We work out the appropriate b_{ij} for each line. We regard these as two independent samples and test whether they represent samples from populations with the same median (the parallel case) using the Wilcoxon–Mann–Whitney test. An example is given by Marascuilo and McSweeney (1977, Section 11.13); see also Exercise 7.14. The test is possible with 10 or more observations in each sample.

7.3 MONOTONIC REGRESSION

There are many cases where a regression of y on x is clearly nonlinear yet it has a monotone property in the sense that as x increases the mean value of y always increases (or at least never decreases), or as x increases the mean y always decreases (or at least never increases).

Conover (1980, p. 274) cites an example where x is the amount of sugar added to standard-size jars of grape juice and y the time taken to complete fermentation. It is well established that increasing amounts of sugar speed up fermentation on average although the rate of increase is by no means constant per unit of additional sugar, which would be a requirement for linear regression.

Similarly, with many learning processes additional tuition will speed up the average rate at which a task can be completed. Again, the relation between tuition time and performance time is usually not linear but will be monotonic. This does not however mean that pairs of individual x, y values will exhibit monotonicity, since there will be variations from average for individuals.

A moment's reflection shows that if we have a monotone regression and transform data to ranks, the ranks exhibit a linear regression. In a few very fortunate cases the ranks will lie exactly in order or inverse order giving Spearman's ρ or Kendall's τ the value $+1$ or -1. The regression for ranks will then be a straight line with slope 1 or -1. In practice random variation usually results in a non-exact match of ranks. Then we may fit by least squares (or by Theil's method) a regression of s on r where r, s are the ranks of the observed x, y. Back-transformation from the estimated $\tilde{s}_i$ corresponding to each r_i gives an expected $\tilde{y}$ corresponding to an observed x. There are many variants on such nonparametric methods of monotone regression. We illustrate the basic ideas by just one example, which presents a modification of a method given by Conover (1980, Section 5.6) and is slightly more straightforward.

Example 7.9

The problem. Fifteen subjects are given varying periods of instruction (x hours) on carrying out a machine operation. After completion of instruction they are tested and the time (y minutes) each takes to complete the machine operation is noted. Fit a monotone regression to the data.

x	0	0	1	1.6	3	4	4	5	6.5	8	8	10	12	12.6	14
y	18.4	17.4	16.2	16.4	14.4	10.5	11.2	10.8	9.0	8.4	7.0	7.2	6.6	5.0	5.6

Formulation and assumptions. The data are already ordered by the x's. We form paired ranks (r, s) (using mid-ranks for ties) and compute the least squares regression of s on r. Writing the regression equation $s = a + br$ for each given r_i we calculate $\hat{s}_i = a + br_i$. We obtain estimates $\hat{y}_i$ of the mean y corresponding to each observed x using, if necessary, linear interpolation between adjacent rankings in the manner described in the procedure. These are joined by straight-line segments for adjacent x values, and the joined segments represent the monotone regression function.

Procedure. The complete procedure is set out in Table 7.4. The first two columns are the data, the columns headed r, s give ranks for these data. The ranks are used to compute the regression coefficients a, b for the linear regression of s on r using least squares. The regression equation is then used to

Table 7.4 Calculating a monotone regression

x	y	r	s	$\hat{s}$	$\hat{y}$
0	18.4	1.5	15	14.38	17.78
0	17.4	1.5	14	14.38	17.78
1	16.2	3	12	12.91	16.38
1.6	16.4	4	13	11.93	16.07
3	14.4	5	11	10.95	14.24
4	11.2	6.5	10	9.47	10.99
4	10.5	6.5	8	9.47	10.99
5	10.8	8	9	8.00	10.50
6.5	9.0	9	7	7.02	9.03
8	8.4	10.5	6	5.55	7.86
8	7.0	10.5	4	5.55	7.86
10	7.2	12	5	4.07	7.01
12	6.6	13	3	3.09	6.64
12.6	5.0	14	1	2.11	5.71
14	5.6	15	2	1.13	5.08

compute entries in the column headed $\hat{s}$ using the corresponding x ranks from the column headed r. Finally $\hat{y}$ is calculated for each $\hat{s}$ by linear interpolation in a manner described below.

The regression of s on r used to obtain $\hat{s}$ is

$$\hat{s} = 15.856 - 0.982r$$

This was computed by the standard least squares method. Using a pocket calculator labour is saved by noting that for ranks

$$b = \frac{\sum r_i s_i - C}{\sum r_i^2 - C}$$

where $C = \frac{1}{4}n(n+1)^2$. Also $a = \frac{1}{2}(1-b)(n+1)$.

To calculate, for example, $\hat{s}_6$ corresponding to the sixth observation in the tabular order, we have $r_6 = 6.5$, whence $\hat{s} = 15.856 - 0.982 \times 6.5 = 9.47$. The remaining entries in the penultimate column are calculated in a similar way.

To calculate entries in the last column we use linear interpolation between adjacent ranks. For example, the rank $\hat{s}_3$ is 12.91, which lies between observation ranks 12 and 13 for s. Now 12.91 represents 91/100 of the interval from 12 to 13, so to get $\hat{y}$ corresponding to this rank we calculate 91/100 of the difference between the y value with rank 12 and the y value with rank 13 and add that to the y value for rank 12. The relevant y values are respectively 16.2 and 16.4. The difference is 0.2 and $0.2 \times (91/100) = 0.182$. Adding this to 16.2 and rounding to two decimal places gives $\hat{y} = 16.38$.

A general formula for linear interpolation between neighbouring assigned y ranks (which need not be integers if mid-ranks are used for ties) is obtained as follows. Let s_i and s_j be respectively the closest 'observation' ranks above and below the relevant $\hat{s}_k$. If y_i, y_j are the observations ranked s_i, s_j then

$$\hat{y}_k = y_j + \frac{s_k - s_j}{s_i - s_j}(y_i - y_j)$$

This formula may be used to calculate all $\hat{y}$. For example, $\hat{s}_{13} = 3.09$ lies between $s_{13} = 3$ and $s_{11} = 4$. Also $y_{13} = 6.6$ and $y_{11} = 7.0$, whence

$$\hat{y}_{13} = 6.6 + [(3.09 - 3)/(4 - 3)] \times (7.0 - 6.6) = 6.636$$

In Table 7.4 this value is rounded to 6.64.

Since we use linear interpolation between corresponding ordered x values it is appropriate to obtain our estimated monotonic regression by plotting each $(x_i, \hat{y}_i)$ and joining points corresponding to adjacent distinct x values by straight-line segments.

Conclusion. The monotone regression is represented by the line segments shown in Figure 7.7.

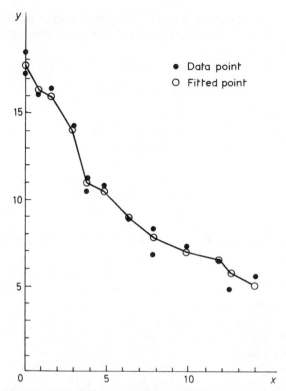

Figure 7.7 Piecewise monotonic regression for data in Table 7.4.

Comments. 1. Sometimes the lowest or highest $\hat{s}_i$ is less than the least s_i or greater than the greatest s_i; then one cannot interpolate between successive ranks, so no $\hat{y}$ can be calculated. The monotonic regression function described here cannot be extrapolated.

2. To determine $\hat{y}_0$ corresponding to x_0 lying between two observed x_i one may read this from the fitted segmented regression or equivalently estimate by linear interpolation using a rank r_0 corresponding to x_0. The regression equation is then used to calculate the corresponding $\hat{s}_0$ and $\hat{y}_0$ is obtained by interpolation as above.

3. In situations such as that described one feels intuitively that a piecewise curve is not satisfactory and that logically a smooth curve with continuous derivative would be more realistic. This is so, and the method given above is very elementary. It is largely of interest as a demonstration of the use of rank transformations and is more objective than fitting a curve by eye. Recently there have been sophisticated developments in nonparametric regression involving the concept of splines. Both the theory and practice of these methods require a basic training in numerical analysis and statistical practice, or

availability of a sound computer program to carry out the computation. A technical account is given by Silverman (1985) and an application is described by Silverman and Wood (1987) and we refer to further work in this area in Section 10.3.2.

4. If any $\hat{s}$ corresponds to an s_i for some observation, the corresponding $\hat{y}$ is the observed y with that rank.

7.4 MULTIVARIATE DATA

We often observe not just two variables but three or more for each experimental unit. Multiple and partial correlation and regression provide parametric methods of dealing with these problems. Partial correlation coefficients corresponding to Kendall's τ are discussed by Conover (1980, Section 5.4). Various extensions of Theil's method to multiple regression have been considered; in their simulation study Hussain and Sprent (1983) found the most satisfactory of those they examined was one proposed by Agee and Turner (1979). It is mathematically sophisticated, requiring orthogonalization of the data and an iterative process using repeated applications of Theil's method that is only feasible with a suitable computer program. The interested reader who wishes to write, or has access to, a suitable computer program should refer to the paper by Agee and Turner for details. Alternatively one of the robust methods outlined in Chapter 10 might be used; again a suitable computer program is required.

7.4.1 Kendall's coefficient of concordance

We indicated in Section 6.5.1 that the Friedman test may be applied to rankings of objects (different varieties of raspberries, placings in a beauty contest, rankings of candidates by different examiners) to test whether there is evidence of consistency or agreement between those making the rankings. Although Kendall originally used a function of the Friedman statistic and tabulated some small-sample critical levels and developed his test independently of Friedman's, there is no reason why one should not use the Friedman test and regard it as a test of the hypothesis H_0: the rankings are effectively random, against H_1: there is evidence of consistency in rankings. Kendall regarded his coefficient of concordance as an extension of the concept of correlation to more than two rankings. Note, however, that whereas in the bivariate case we may use Kendall's τ as a measure of agreement (positive correlation) or disagreement (negative correlation), concordance, whether measured by Kendall's original statistic or the Friedman modification, is one-sided in the sense that rejection of the null hypothesis indicates a consistency between, for example, judges in a beauty contest. If we have four judges A, B, C, D ranking five objects I, II, III, IV, V and they award the ranks below, we would not reject H_0; the Friedman statistic (6.6) then takes the value zero.

Object	Judge			
	A	B	C	D
I	1	5	1	5
II	2	4	2	4
III	3	3	3	3
IV	4	2	4	2
V	5	1	5	1

This is a situation where there is complete agreement between judges A and C and between judges B and D; but the latter pair are completely at odds with judges A and C. The Kendall or Friedman statistics do not detect such patterns.

7.5 FIELDS OF APPLICATION

We look first at situations where one may wish to calculate rank correlations.

Political science

Leaders of political parties may be asked to rank certain problems facing Parliament in order of importance. If there are two leaders, rank correlations will be of interest. If leaders of more than two parties are involved the coefficient of concordance may be appropriate, or we may prefer to look at pairwise rank correlations, for leaders of very different political colour may well tend to reverse, or partly reverse, rankings.

Psychology

A psychologist may show 12 different photographs separately to a twin brother and sister and ask each to rank them in order of the pleasure they give. When the ranks given by brother and sister to each photograph have been recorded, either Kendall's τ or Spearman's ρ may be calculated to see if there is evidence of consistency in the preferences.

Business studies

A firm of market research consultants list a number of aspects of good sales practice such as consistent quality of goods, reasonable guarantees, keen pricing, quick after-sales service, clear operating instructions, etc. They ask a manufacturers' association and also a consumers' association each to rank these characteristics in order of importance. A rank correlation coefficient will establish whether there is any measure of agreement between manufacturer and consumer views on relative importance.

Personnel management

A personnel officer may rank 12 salesmen on the basis of total sales by each over a given period. His boss suggests he should also rank them on the basis of numbers of customer complaints received about each. A rank correlation coefficient could be used to indicate how well the rank scores agree. A Pearson product moment coefficient could also be calculated if we had figures for actual sales and for precise numbers of customer complaints for each salesman.

Horticulture

Leaf samples may be taken from each of 20 trees and the magnesium and calcium contents determined by chemical analysis. A Pearson coefficient might be used to see if levels of the two substances are related, but this coefficient can be distorted if one or two trees have levels of these chemicals very different from the others; such influential observations are not uncommon in practice. A rank correlation coefficient may give a better picture of the correlation. If a third chemical, say cobalt, is also of interest, a coefficient of concordance might be appropriate.

Regression is one of the most widely used of all statistical techniques. In all fields where least squares regression is used, Theil's method provides a robust nonparametric alternative in the bivariate case. We mentioned briefly some multiple regression alternatives in Section 7.4 and others are mentioned in Sections 9.1.3 and 10.3.2.

EXERCISES

7.1 In Table 7.1 ages at death are ordered by year of death. Use the Cox–Stuart trend test (Section 2.1.6) to test for a time trend in lifespans.

7.2 A china manufacturer is investigating market response to seven designs of dinner sets. The main outlets are the British and American market. To get some idea of preference on the two markets a survey of 100 British housewives and 100 American housewives is carried out to rank the designs in order of preference from 1 for favourite to 7 for least desirable. For each country the 100 ranks scores for each design are totalled. The design with the lowest total is assigned rank 1, that with the next lowest total rank 2, and so on. The overall rankings for each country are as follows:

Design	A	B	C	D	E	F	G
British rank	1	2	3	4	5	6	7
American rank	3	4	1	5	2	7	6

Calculate Spearman's ρ and Kendall's τ for these data. Is there evidence of a direct association in order of preference in the two countries?

7.3 The manufacturer in Exercise 7.2 later decides to assess preferences in the Canadian and Australian markets by the same methods. The following rankings are obtained:

Design	A	B	C	D	E	F	G
Canadian rank	5	3	2	4	1	6	7
Australian rank	3	1	4	2	7	6	5

Calculate Spearman's ρ and Kendall's τ for these data. Is there evidence of a direct association in order of preference in the two countries?

7.4 Perform an appropriate analysis of the rank data for all four countries in Exercises 7.2 and 7.3 to determine whether there is evidence of any overall agreement about ranks. Comment on the practical implications of your result.

7.5 In a pharmacological experiment involving β-blocking agents, Sweeting (1982) recorded for a control group of dogs cardiac oxygen consumption (MVO) and left ventricular pressure (LVP). Calculate Spearman's ρ and Kendall's τ. Is there evidence of correlation?

Dog	A	B	C	D	E	F	G
MVO	78	92	116	90	106	78	99
LVP	32	33	45	30	38	24	44

7.6 Bardsley and Chambers (1984) give numbers of beef cattle and sheep on 19 farms in a region. Is there evidence of correlation?

Cattle	41	0	42	15	47	0	0	0	56	67	707
Sheep	4716	4605	4951	2745	6592	8934	9165	5917	2618	1105	150

Cattle	368	231	104	132	200	172	146	0
Sheep	2005	3222	7150	8658	6304	1800	5270	1537

7.7 We give below the numbers of rotten oranges, y, in ten randomly selected boxes from a large consignment after they have been kept in storage for a stated number of days, x. Use Theil's method to fit a straight line to these data.

Days storage (x)	3	5	8	11	15	18	20	25	27	30
No. rotten (y)		2	4	7	10	17	23	29	45	59	73

Plot the data, and your fitted line, on graph paper. Does a straight-line fit seem reasonable?

7.8 The data below give the modal lengths (y cm) of samples of Greenland turbot of various ages (x yr) based on data given by Kimura and Chikuni (1987). Fit a monotone regression to these data.

Age (x)	4	5	6	7	8	9	10	11	12
Length (y)	40	45	51	55	60	67	68	65	71

Age (x)	13	14	15	16	17	18	19	20
Length (y)	74	76	76	78	83	82	85	89

7.9 Comment critically on and explain the meaning of the following statement. Given a set of n observations (x_i, y_i) where the x_i are equally spaced, the Cox–Stuart test for trend in the y_i is essentially equivalent to testing whether the abbreviated Theil estimator is or is not consistent with zero slope.

7.10 Mattingley (1987) quotes data based on the US census of agriculture which gives at approximately 10-year intervals from 1920 to 1980 the percentage of farms with tractors and farms with horses.

Percentage with tractors	9.2	30.9	51.8	72.7	89.9	88.7	90.2
Percentage with horses	91.8	88.0	80.6	43.6	16.7	14.4	10.5

Fit a monotone regression of horse percentages on tractor percentages. Explain why it would be pointless to fit a linear regression to these data.

7.11 Paul (1979) discusses marks awarded by 85 different examiners to each of 10 scripts. The marks awarded by six of these examiners were as follows:

	Script									
Examiner	1	2	3	4	5	6	7	8	9	10
1	22	30	27	30	28	28	28	28	36	29
2	20	28	25	29	28	25	29	34	40	30
3	22	28	29	28	25	29	33	29	33	27
4	24	29	30	28	29	27	30	30	34	30
5	30	41	37	41	34	32	35	29	42	34
6	27	27	32	33	33	23	36	22	42	29

Use rank tests to determine: (i) whether the examiners show reasonable agreement on order of merit of the scripts, and (ii) whether some examiners tend to give consistently higher or lower marks than others.

7.12 Wahrendorf, Becher and Brown (1987) give data for doses in μg of a tumour-inducing compound B(a)P on the percentage of animals developing tumours at each dose. Fit a monotone regression to these data. Does the fit seem reasonable? Comment critically on the usefulness (if any) of this regression.

Dose	0.05	0.5	5	20	50	100	300
Percentage tumours	1.0	0.0	4.9	44.2	30.0	86.5	56.9

7.13 Gat and Nissenbaum (1976) give ammonia concentrations at various depths in the Dead Sea. Fit a linear regression for concentration on depth by Theil's method and obtain a 95% confidence limit for β.

Depth (m)	25	50	100	150	155
N-NH$_3$(mg/l)	6.13	5.51	6.18	6.70	7.22

Depth (m)	187	200	237	287	290	300
N-NH$_3$(mg/l)	7.28	7.22	7.48	7.38	7.38	7.64

7.14 Katti (1965) gives the following data for weight of food eaten (x) and gains in weight (y) for 10 pigs given one type of food and for 10 given a second type of food. Use Theil's abbreviated method to fit linear regressions to each and test whether the true slopes may be accepted as equal.

First feed
x 575 585 628 632 637 638 661 674 694 713
y 130 146 156 164 158 151 159 165 167 170

Second feed
x 625 646 651 678 710 722 728 754 763 831
y 142 164 149 160 184 173 193 189 200 201

8
Counts and categories

8.1 CATEGORICAL DATA

Data often consist of counts of numbers of objects with given attributes or belonging to given categories arranged into one-, two-, three- or even higher-dimensional tables, usually referred to as one-, two-, three-way contingency tables. Each dimension or 'way' corresponds to a classification into categories representing one attribute.

8.1.1 One-way tables

Table 8.1 is a one-way table of numbers of cities with over 1 million inhabitants in six different countries (latest published data, 1987). Categorization is by country.

Table 8.1 Numbers of cities with populations exceeding 1 million

Country	Brazil	W. Germany	India	Gt Britain	Australia	USA
No. cities	6	3	10	2	4	6

The data in Table 8.1 are unlikely to be relevant to any statistical analysis. One could go through the motions of pretending the data were a random sample of countries and obtain an estimate of the median or mean number of cities per country with population over 1 million, but the result would not be informative or valid because the data were not randomly selected. I selected them recently on subjective criteria for another purpose. If we had total populations for each of these countries we might then study the relationship between numbers of large cities and total population; here again lack of randomness makes it questionable to what 'population' (in the statistical sense) any inferences would apply.

Table 8.2 shows the numbers of times each of the digits 0 to 9 occurs as the penultimate digit in one column of the 1986 telephone directory for Tayside and North Fife, Scotland; e.g. the penultimate digit in the number 32759 is 5.

These data might be accepted as an effectively (though not a truly) random sample of entries in a telephone directory, and one may want to test a hypothesis that penultimate digits are uniformly distributed; i.e. that any digit is equally likely to occur. The test is not really nonparametric as it is a test

Table 8.2 Numbers of occurrences of digits in penultimate position of telephone numbers

Digit	0	1	2	3	4	5	6	7	8	9
Frequency	13	12	10	13	11	16	12	13	17	13

about a parameter, p, that specifies the frequency of occurrence of any digit. It is a **goodness of fit test** that we give in Section 8.4 and the justification for regarding it as distribution-free is that it can be used to test hypotheses about any discrete distribution.

We may wish to perform other and undoubtedly nonparametric tests with the data in Table 8.2. Are odd digits more common than even digits? Does the frequency of digits increase as the digit value proceeds from 0 to 9? The Cox–Stuart trend test would be appropriate to test the latter, although the sequence is necessarily short and could only provide a meaningful critical region for a one-tail test.

8.1.2 Two-way tables

Two-way tables that are similar in format may differ greatly in logical status. We illustrate some of those differences by examples. If, in addition to the data in Table 8.2, we also have a sample from some other directory page of frequencies of the final digit in a column of telephone numbers, we may present the combined data in a two-way table. One classification is by digit numbers and the other by the categories penultimate and final digit. Table 8.3 was obtained this way.

Table 8.3 Digit frequencies in penultimate and final positions

Digit	0	1	2	3	4	5	6	7	8	9
Penultimate	13	12	10	13	11	16	12	13	17	13
Final	14	11	14	8	18	8	11	11	11	14

The data consist of two independent samples. We may ask whether they support a hypothesis that the distributions of digits are the same in the penultimate and in the final digit position. This is a nonparametric and distribution-free situation. We do not specify that the distribution need be uniform or of any other form. We simply ask whether it is reasonable to infer from the data that there is no dependence of the proportion of occurrences of any given digit upon its position – penultimate or final. The test is of interest because while it might be reasonable to suppose penultimate digits are effectively random in an alphabetic list of telephone subscribers, this may not be true for final digits since large institutions often have a number of exchange lines and only one leading number is listed in a directory. On some exchanges the practice is to list a number such as 241701 with last digit 1 when a business subscriber also has the numbers 241702, 241703, 241704,... which are not listed.

We met a table with two samples in Example 5.7. The data for that example are reproduced in Table 8.4.

Table 8.4 Page number distribution for two types of book

Pages	< 100	100–199	200–299	300–399	400–499	500 +
Statistics books	0	4	5	8	3	4
General books	5	4	5	1	0	1

In Example 5.7 we were interested primarily in a test for distribution medians. We might ask the wider question whether the classifications into numbers of pages (i.e. the proportions in each of the categories < 100, 100–199, etc.) are independent of the type of book (statistical or general). One of the classifications (that by number of pages) is ordinal. The other is categorical with no implied order.

We also meet two-way tables where we have one sample but characterize individual sample units with respect to two or more unique characteristics. Schoolchildren are typically graded in examinations into passes with grades, say, A, B, C, D, E, where these are ordered, e.g. a grade A performance is better than grade B, grade B better than C, etc. We might take a sample of pupils taking the examination and count not only the numbers in each grade but also note the sex of each candidate, leading to a two-way table like Table 8.5. The row totals are given to aid the discussion below.

Table 8.5 Examination results by grade and sex

Grade	A	B	C	D	E	Total
Male	7	23	31	19	12	92
Female	3	14	29	18	10	74

Only after we have selected our sample do we know how many of each sex are in the whole sample and in each grade. We may use such data to see if there is dependence between grade and sex; i.e. whether the proportions in each grade in the population are independent of sex. We could also have obtained data like that in Table 8.5 by deciding beforehand that we wanted a sample of 92 boys and a sample of 74 girls and selected samples of that size independently for each sex. Although we might expect much the same story about dependence to emerge from samples selected either way, the logic of what we do is different. In the second case we fix the row totals beforehand at 92, 74. If, on the other hand we select random samples of 166 pupils, we will not in every sample get 92 boys and 74 girls. If we repeated the experiment we might next time get 89 boys and 77 girls, the next time 81 boys and 85 girls. We shall see in Section 8.2 that if we base tests on randomization of possible outcomes for any given sample, the relevant test criteria differ. However, in the two cases just described that difference becomes negligible for moderate sample sizes and we may use the same large-sample test irrespective of whether selection was by:

1. fixing only the total number of pupils to be sampled and then noting numbers in each grade/sex combination or
2. fixing also the numbers of each sex and then noting grades for those selected for each sex.

In one situation we fix only the grand total; in the other the grand total and row totals. It is also possible to fix the grand total and column totals; this is equivalent to an interchange of rows and columns. There is a further case in which we fix both row and column totals.

The large-sample approximate test for independence between the row and column classifications is the same in all these situations and is essentially a test which is conditional upon the observed row and column totals.

In the following example both row and column totals are fixed. We may be interested in whether a firm tends to reply by first-class mail to incoming mail that has been despatched to them first class or whether there is no association between the grades of incoming and outgoing mail. We suppose further that their policy is to despatch two-thirds of their mail first class and one-third second class. To see whether there is any dependence of the outgoing classification on the incoming classification we record for one day's incoming mail 27 first-class and 21 second-class letters. These fix marginal totals for the incoming mail as it is determined without the control of the firm. Because of their policy, 32 replies will be sent first class and 16 second class; these fix the other marginal totals. The incoming and outgoing mail pattern for the 48 letters received and the replies to each might be that given in Table 8.6.

In introductory statistics courses it is common to assume that row and column totals are both fixed. As our examples have indicated this may not always be so. Though it may not be strictly true, it is often not unreasonable to assume that under an appropriate null hypothesis they may be regarded as fixed. For example, if a new drug is to be tested for a disease where there was previously no treatment – it being a matter of chance whether or not a patient recovered – it may be decided to divided 19 available patients into a group of 10 who are to receive the new drug and the remaining 9 who will not be treated. Under the null hypothesis that the drug has no effect we can consider ourselves as being in a position that fate may already have determined who is to recover and who is not to recover. This in effect fixes the other marginal totals of numbers who recover or fail to recover. Random allocation of patients then gives a two-way table (say, row classification by treated/untreated and column classification by recovered/not recovered) with fixed marginal totals. If the drug is effective this should be indicated by an appropriate test as to whether the data are likely to arise by randomization under the null hypothesis. Indeed we test conditionally upon the given marginal totals. This and the fact that many large-sample tests are equivalent whether or not marginal totals are controlled by the experimenter justify a commonly made assumption that we may proceed as if they were in so far as testing hypotheses of independence of row and column classifications are concerned. For the simple 2×2 table, the case for proceeding this way is well argued by Yates (1984).

We met a different situation in Section 4.2 where we considered essentially one sample but two observations on each unit in that sample. The example we gave involved success and failure on two climbs and the data are reproduced in Table 8.7.

Table 8.6 Classification of first- and second-class mail

| | | Inward class | | |
		First	Second	Total
Outward class	First	26	6	32
	Second	1	15	16
	Totals	27	21	48

Table 8.7 Performance of mountaineers attempting two climbs

| | | First climb | |
		Success	Failure
Second climb	Success	73	14
	Failure	9	12

In Example 4.3 we proposed the McNemar test for the data. The tests to be described in Sections 8.2.1 and 8.2.2 are not appropriate here, for we are no longer interested in the hypothesis of independence between the two categories, but rather in comparing matched pair responses with a large number of ties within pairs. The appropriate large-sample test is given in Section 8.2.5 (it is equivalent to the related Cochran Q test, Section 6.5.4).

8.1.3 Multi-way tables

While two-way tables are easily represented on paper, three- or more-way tables are best represented by sub-tables and more than one such representation are always possible. For example, in a study of coronary heart disease patients may be categorized into those who have and have not coronary heart disease (CHD) (classification 1); then they may be listed in three categories of cholesterol level (classification 2) and into five blood pressure levels (classification 3). One way to set out such data is given in Table 8.8; it is convenient for a later analysis to include the indicated marginal totals.

Table 8.8 Categorization of 200 patients according to three factors

| | | Blood pressure grade | | | | | Total |
		I	II	III	IV	V	
	Cholesterol						
	A	2	3	1	0	4	10
CHD	B	2	1	5	3	0	11
	C	4	7	8	6	2	27
	Total (1)	8	11	14	9	6	48
	A	16	14	11	8	6	55
No CHD	B	22	18	5	3	2	50
	C	15	13	10	5	1	44
	Total (2)	53	45	26	16	9	149
	Total (1 + 2)	61	56	40	25	15	197

The reader should experiment with other logical arrangements of these data. Although tests for complete independence can be extended from two- to multi-way tables and pairwise tests of independence may be used, such analyses are usually insufficient (and inefficient) and we are often interested in testing more elaborate hypotheses, or even a series of hypotheses – a matter to which we return in Section 8.3.

8.2 TESTS FOR INDEPENDENCE IN TWO-WAY TABLES

Before presenting one exact test and a large-sample approximation, we first indicate by simple illustrations with 2×2 tables that the randomization theory is different for models with different categories of fixed marginal totals; hence, despite the conditional and the asymptotic, or large-sample, theory agreeing, exact tests differ and are complicated.

8.2.1 Randomization tests and marginal constraints

If we wished to compare two treatments and had 15 patients and allocated them at random in groups of 7 and 8 to each treatment and each were equally effective (our null hypothesis) we should expect about the same proportion of successes with each. A reasonable model is that we are observing two independent samples from binomial distributions each with the same p. The theory is well known. Clearly with such an assumption if we consider a two-way table with success and failure as outcomes and the indicated sample sizes for treatments A and B we may express the outcomes in the form:

	Success	Failure	Total
Treatment A	x	$7 - x$	7
Treatment B	y	$8 - y$	8

In this case there are 72 possible outcomes involving any of the values 0 to 7 for x combined with any of 0 to 8 for y. To compare some of these we shall use the shorthand pattern

$$a \quad b$$
$$c \quad d$$

to represent entries in the body of the table (each position is often called a **cell**), i.e. $a = x$, $b = 7 - x, c = y$, $d = 8 - y$ in the above example. In this notation clearly an outcome like

$$4 \quad 3$$
$$4 \quad 4$$

favours the null hypothesis that the probabilities of success are the same for each sample. So would an outcome like

$$2 \quad 5$$
$$3 \quad 5$$

but intuitively we would feel the equality occurred for a lower value of p. An outcome like

$$0 \quad 7$$
$$8 \quad 0$$

suggests very different values of p for the two samples; perhaps $p = 0$ for the first and $p = 1$ for the second. A problem in working out an exact test is that the probability of achieving each of the 72 configurations under the null hypothesis depends on the usually unknown p, a matter discussed in detail by Conover (1980, Section 4.1), who shows that difficulties arise even in the trivial case where the two samples are each of size 2, when there are only 9 different contingency tables. Fortunately these difficulties are surmountable by using the argument that under the null hypothesis that p is the same for each sample we may estimate it by combining the data for both samples; effectively we then test conditional upon the marginal totals, an argument emphasized by Yates (1984), although implicit in work by Fisher, Yates and others 50 years earlier.

The situation is even more open for a single sample categorized by two characteristics (see e.g. Table 8.5) where only the grand total is fixed. There are clearly many different tables. Conover (1980, p. 161) shows that for a 2×2 table with a fixed total of 4 there are 35 contingency tables compared with 9 for the two-sample case with one set of marginal totals fixed at 2 and 2. At the other extreme, if both row and column marginal totals are fixed, then in the case of a total of 4 if the fixed marginals are 3, 1, for rows and 2, 2 for columns there are clearly only two possible tables, namely

$$
\begin{matrix} 2 & 1 \\ 0 & 1 \end{matrix} \quad \text{and} \quad \begin{matrix} 1 & 2 \\ 1 & 0 \end{matrix}
$$

For larger totals there are more possible tables. For example, with $N = 15$, row totals 7 and 8, column totals 4 and 11 it is easily verified that possible contingency tables are

$$
\begin{matrix} 4 & 3 \\ 0 & 8 \end{matrix} \quad \begin{matrix} 3 & 4 \\ 1 & 7 \end{matrix} \quad \begin{matrix} 2 & 5 \\ 2 & 6 \end{matrix} \quad \begin{matrix} 1 & 6 \\ 3 & 5 \end{matrix} \quad \begin{matrix} 0 & 7 \\ 4 & 4 \end{matrix}
$$

More generally, if we denote the cell entries by a, b, c, d such that the contingency table is

$$
\begin{matrix} a & b \\ c & d \end{matrix}
$$

where $a + b$, $c + d$, $a + c$, $b + d$ are fixed, Fisher (1922) showed that under the null hypothesis of no association the probability of that configuration is

$$
\frac{(a + b)!(c + d)!(a + c)!(b + d)!}{(a + b + c + d)!\,a!b!c!d!} \tag{8.1}
$$

where $x!$ is the product of all integers between 1 and x, e.g. $4! = 1 \times 2 \times 3 \times 4$.

Tedious calculation show that the probabilities of each of the configurations under H_0 given by (8.1) are, for our numerical example,

$$
0.0256, \ 0.2051, \ 0.4308, \ 0.2872, \ 0.0513
$$

These add, as they should, to 1.

For a test at the nominal 5% level (exact level 2.56%) the critical region is given by the first configuration alone. The test is often called Fisher's exact test (or sometimes the Fisher–Irwin test) and is applicable for inferences conditional on fixed marginal totals which, as we have indicated, may be appropriate under a null hypothesis when we have, for instance, two independent samples and wish to test for a common p.

8.2.2 A large-sample test for independence

In a large-sample test for independence in two-way tables, whether or not marginal totals are formally fixed, we customarily estimate expected numbers in each cell under the assumption of independence with the aid of the actual marginal totals, i.e. conditional upon those totals.

We consider in general a contingency table with r rows and c columns; often called an $r \times c$ table. We denote the observation (or count) in cell ij (in the ith row and jth column) by n_{ij}; the ith row total by n_{i+}, $i = 1, 2, \ldots, r$ and the jth column total by n_{+j}, $j = 1, 2, \ldots, c$ and the grand total by N (i.e. $N = \sum_i n_{i+} = \sum_j n_{+j}$). Under H_0: the classifications are independent, the expected number in cell ij is $e_{ij} = (n_{i+})(n_{+j})/N$. For example, in the 2×2 contingency table with observations:

	R_1	R_2	Total
C_1	6	3	9
C_2	5	1	6
Total	11	4	15

the expected number in the first cell (row 1, column 1) is $e_{11} = (9 \times 11)/15 = 6.6$. For row 1, column 2, $e_{12} = (9 \times 4)/15 = 2.4$. Proceeding thus we find the expected numbers in each cell are

$$6.6 \quad 2.4$$
$$4.4 \quad 1.6$$

The row and column totals for observed and for expected numbers are the same, thus in a 2×2 table it is only necessary to calculate one expected number *ab initio* – the others are obtained by subtraction from the appropriate marginal totals. In the case of an $r \times c$ table, only the expectations in the first $(r-1)$ rows and $(c-1)$ columns need be obtained *ab initio*. The others then follow automatically to ensure correct marginal totals.

The null hypothesis of independence is supported if observed and expected numbers are close. This is the basis of the large-sample test where one test statistic is calculated as follows:

1. For each cell calculate

$$\frac{(n_{ij} - e_{ij})^2}{e_{ij}}$$

2. Add the result in (1) above for each cell, whence

$$T = \sum_{i,j} \frac{(n_{ij} - e_{ij})^2}{e_{ij}} \tag{8.2}$$

3. We compare T with critical values of the chi-squared distribution with $(r-1)(c-1)$ degrees of freedom and reject H_0 if T exceeds the critical value at the chosen significance level.

An equivalent form of (8.2) computationally more convenient is

$$T = \sum_{i,j} \frac{n_{ij}^2}{e_{ij}} - N \qquad (8.3)$$

Example 8.1

The problem. A company has two manufacturing plants each turning out the same product. Each item produced is graded into one of three quality grades by quality-control inspectors. A sample of 50 is taken from the output of plant A and a sample of 40 from the output of plant B. The numbers in each of the three grades super, standard, second are as follows:

	Super	Standard	Second	Total
Factory A	2	41	7	50
Factory B	5	32	3	40
Total	7	73	10	90

Is it reasonable to suppose proportions in each quality grade are the same for each factory?

Formulation and assumptions. We have a 2 × 3 table with fixed row totals. Can proportions in each category from each factory reasonably be supposed the same? We use the large-sample version of the test and the chi-squared statistic with $(r-1)(c-1) = 2$ degrees of freedom.

Procedure. We first calculate the expected numbers in each cell, e.g. that in row 1, column 1 is $e_{11} = (50 \times 7)/90 = 3.89$. It suffices to calculate to two decimal places. Similarly $e_{12} = (73 \times 50)/90 = 40.56$. The remaining entries may be obtained by subtraction, although as a check on arithmetic, it pays to get one more directly. There may be slight rounding differences in the last decimal place. The expected numbers are

$$3.89 \quad 40.56 \quad 5.56$$
$$3.11 \quad 32.44 \quad 4.44$$

Using (8.2) or (8.3) we find $T = 2.918$. From chi-squared tables the critical value with 2 d.f. when testing at the 5% level is 5.99.

Conclusion. Our T is well-below the critical value; we do not reject the null hypothesis of independence.

Comments. 1. The large-sample test works well even if there are a few cells with low expectation. What is meant by 'few' and 'low' in this context is somewhat controversial. A widespread rule of thumb is that all e_{ij} should be 5 or more, but many writers argue that this can reasonably be relaxed for a small proportion of cells. The large-sample test may be misleading in sparse tables where there are numerous empty cells.

2. With small numbers in some cells it may be advantageous to combine adjacent rows or columns to give reasonable numbers. This loses some degrees of freedom, but there is often little loss in power.

In 2 × 2 tables a continuity correction is often recommended. Using this, to form T we calculate for each cell $(|n_{ij} - e_{ij}| - \frac{1}{2})^2/e_{ij}$. This gives a more conservative test but the adjustment is of little importance unless T is close to a critical value and it is important not to overestimate significance. In such situations one might be better to use Fisher's exact test. The continuity correction is only appropriate to 2 × 2 tables.

8.2.3 An alternative test statistic

An alternative statistic based on maximum likelihood (a method of estimation with many common-sense desirable properties but unfortunately not a monopoly of virtues) is

$$T_1 = 2\sum_{i,j} n_{ij} \ln(n_{ij}/e_{ij}) \qquad (8.4)$$

Here ln is used to denote the natural logarithm or logarithm to the base e. Many pocket calculators have a key to calculate this. Like T in (8.2), T_1 has asymptotically a chi-squared distribution with $(r-1)(c-1)$ degrees of freedom under the null hypothesis of independence. In practice the calculated values of T and T_1 often differ markedly when alternative hypotheses hold, but conclusions about significance levels using either test are usually similar unless the sample size is small. Because of desirable additivity properties, particularly in the more sophisticated testing procedures developed in Section 8.3, T_1 is often preferred to T.

For the data in Example 8.1 we find $T_1 = 2.97$, leading to the same conclusion as we get with $T = 2.18$.

8.2.4 The median test

The median test introduced in Section 5.1.1 extends to several samples in a rather obvious manner that leads to a large-sample test involving in the case of k samples a 2 × k contingency table. We first determine M, the median for the combined sample data. Then for each sample we determine the numbers of observations a_i, b_i, $i = 1, 2, \ldots, k$ that are above and below the median respectively. We exclude from our counts sample values equalling M. We may set out our results in the form:

Sample	1	2	3	...	k
No. below M	b_1	b_2	b_3		b_k
No. above M	a_1	a_2	a_3		a_k

Under H_0: the samples have identical distributions, we expect a_i, b_i to be nearly equal and if no sample values equal M the two row totals will be equal. Even with some values equal to M, these totals are not likely to differ markedly. For each sample $a_i + b_i = n_i$, the sample size, if no value equals M. The test statistic if T given by (8.2) or (8.3) or T_1 given by (8.4). Although rather different in appearance, the test statistic given for the two-sample case in Section 5.1.1 is equivalent to (8.2).

Example 8.2

The problem. Apply the median test to the number of pages of data in Table 6.1.

Formulation and assumptions. We require the grand median of all 36 observations. If there are no location differences there should be near equality of numbers of observations above and below the median for each sample.

Procedure. The median of all numbers of pages is 245.5, the mean of the 18th and 19th observations 245 and 246. Numbers above and below this value for each sample are:

	Travel	General	Statistics
No. below	3	10	5
No. above	5	6	7

Clearly both row totals are 18 and the expected numbers above or below for each class are respectively 4 for travel, 8 for general, and 6 for statistics, whence T given by (8.2) is easily found to be 1.83.

Conclusion. Since the T value is well below that required for significance with 2 degrees of freedom we do not reject H_0.

Comment. While easy to perform, the median test is generally less efficient than the Kruskal–Wallis test.

8.2.5 McNemar's test

McNemar's test given in Section 4.2 is not a test for independence, but there is a chi-squared approximation to it valid for large samples that is equivalent to the normal approximation to the binomial. For the data in Table 8.7, for example, we effectively test whether 9 successes and 14 failures (or vice versa) are consistent with a binomial distribution with $p = \frac{1}{2}$ and $n = 9 + 14 = 23$. The expected numbers of successes and failures is 11.5 and the chi-squared test statistic is $T = 2.5^2/11.5 + (2.5)^2/11.5 = 1.09$. This is compared with the relevant critical value for chi-squared with 1 degree of freedom. It is easily verified that if we have a success/failure combinations and b failure/success combinations ($a = 9$, $b = 14$ in Table 8.7) then T reduces to

$$T = (a - b)^2/(a + b)$$

In Section 6.5.4 we introduced Cochran's test for comparing several treatments with binary $(0, 1)$ responses. When $t = 2$ it is not difficult to show (see Exercise 8.12) that Cochran's Q reduces to the form of T just given.

8.2.6 Tests of association

The tests of independence in Sections 8.2.2 and 8.2.3 beg the question of how we measure degree of dependence if we reject H_0. The subject is a complex one and some measures of dependence are discussed by Conover (1980, Section 4.4). More sophisticated approaches use the log-linear model.

8.3 THE LOG-LINEAR MODEL

A detailed analysis of three- or more-way contingency tables usually calls for more than simple tests of pairwise independence which may be performed on sub-sets of two-way tables. A large number of different models are possible and modern analytic techniques are often based on the log-linear model. In an elementary text it is appropriate only to give a brief introduction to these models and tests, illustrated by one simple example. The analytic techniques are discrete data analogues to the analysis of variance for continuous data. Readers not already familiar with linear model analysis of variance for experimental designs with factorial treatment structures may find this section difficult to follow, but it is hoped an elementary explanation will give some indication of the power of the log-linear model as an analytic tool. More sophisticated applications of the method need both a broad knowledge of statistics and the availability of suitable computer programs. This short account is primarily to draw attention to a technique not widely known to research workers who are not primarily statisticians, but who carry out much of their own data analysis. Marascuilo and McSweeney (1977, Ch. 9) give further examples of the use of the log-linear model.

The numerical example we use was first discussed by Bartlett (1935) and is a classic example of the analysis of a $2 \times 2 \times 2$ contingency table. The data are survival numbers of plant cuttings in each of 4 batches of 240 cuttings. The first batch consisted of long cuttings planted early, the second of long cuttings planted late, the third of short cuttings planted early and the fourth of short cuttings planted late. The three modes of classification, each with two categories, were (1) length of cutting (long or short), (2) date of planting (early or late) and (3) survival status (dead or alive). Bartlett's data are summarized in Table 8.9.

Table 8.9 Survival status of cuttings under different conditions

Length of cutting	Long		Short	
Time of planting	Early	Late	Early	Late
Alive	156	84	107	31
Dead	84	156	133	209

Visual inspection of Table 8.9 suggests that even if we consider long cuttings only, survival rates are almost certainly not independent of time of planting. This is confirmed with a chi-squared value of 43.2 with 1 degree of freedom for this sub-set. The same would seem to be true for short cuttings but the pattern here looks rather different. The log-linear model provides a way of analysing such differences. As an introduction we first discuss analogous situations for a linear model.

8.3.1 Main effects and interactions in a linear model

We digress to discuss experiments with responses (usually some sort of measurement) on a continuous scale. We consider **factorial treatment structures**. The simplest of these is the so-called 2×2 structure, where we have two factors each at two levels. For example, in a chemical process a reaction may be allowed to proceed for two periods (first factor), e.g. 2 hours or 3 hours; also at each of two temperatures (second factor), say 75°C and 80°C. The response of interest might be the amount of some chemical produced by a given volume of input material using each factor combination. Such an experiment is repeated (or replicated) and the average (or total) output for the same number of replicates for each factor combination is recorded. Apart from random variation in the form of an additive error which, in the analysis of variance model, is assumed to be distributed normally with a mean zero and variance σ^2, the output x_{ij} for the first factor at level i and the second at level j is specified by a linear (additive) model. If the effects of each factor are purely additive we speak of a **no-interaction** model.

Ignoring for the moment random variation, suppose, for instance, we found that operating for 2 hours at 80° increased output by 3 units above what we got by operating for 2 hours at 75° and that operating for 3 hours at 75° increased our output by 8 units above what we got by operating for 2 hours at 75°; then we say we have a no-interaction model if operating for 3 hours at 80° increased output from that for 2 hours at 75° by the sum of these increases, i.e. by $3 + 8 = 11$ units. If our mean output for 2 hours at 75° is X we may summarize these results as in Table 8.10.

Table 8.10 Expected yields in a no-interaction model

| | | Temperature (°C) | |
		75	80
Process time	2	X	$X + 3$
(hr)	3	$X + 8$	$X + 3 + 8$

More generally, let us denote expected yields (values of output apart from random variation) by $\tilde{x}_{ij}$, the tilde being used to distinguish this from an actual yield x_{ij} which contains an additive random variation, or error, component.

In Table 8.10, $\tilde{x}_{11} = X$, $\tilde{x}_{12} = X + 3$, $\tilde{x}_{21} = X + 8$, $\tilde{x}_{22} = X + 11$. Note that here $\tilde{x}_{11} + \tilde{x}_{22} = 2X + 11 = \tilde{x}_{12} + \tilde{x}_{21}$, i.e. the diagonal or 'cross' sums in Table 8.10 are equal. This is a characteristic of a no-interaction model. In many real-life situations we find the effect of increasing both time and temperature is to boost or diminish the effect we get from changing only one factor. This would be the situation if, for example, $\tilde{x}_{22}$ in Table 8.10 had the value $X + 17$ or $X + 2$ rather than $X + 11$. The cross-sums are then no longer equal.

For a 2×2 factorial model with no interaction the key requirement is

$$\tilde{x}_{11} + \tilde{x}_{22} - \tilde{x}_{12} - \tilde{x}_{21} = 0$$

If there is an interaction (as there would be if we replaced the entry for x_{22} in Table 8.10 by $X + 16$) we now find the cross-difference is $X + (X + 16) - (X + 3) - (X = 8) = 5$. This difference of 5 is a measure of interaction between the factors.

In general, an interaction implies

$$\tilde{x}_{22} + \tilde{x}_{11} - \tilde{x}_{12} - \tilde{x}_{21} = I, \qquad \text{where } I \neq 0$$

In the analysis of variance of experimental results the observed means $\bar{x}_{ij}$ for each factor combination provide sample estimates of the unknown $\tilde{x}_{ij}$. Tests based on the F-distribution are used to see if there is a significant interaction between factors, or whether an observed nonzero value of the cross-difference $\bar{x}_{11} + \bar{x}_{22} - \bar{x}_{12} - \bar{x}_{22}$ can reasonably be attributed to random variation.

8.3.2 The independence anology

In a 2×2 contingency table our expected value e_{ij} for cell ij under the hypothesis of independence or no association is $e_{ij} = (n_{i+})(n_{+j})/N$. It immediately follows that

$$e_{11}e_{22} = (n_{1+}n_{+1}n_{2+}n_{+2})/N^2 = e_{12}e_{21}$$

i.e. that the cross or diagonal products are equal, or equivalently

$$\frac{e_{11}e_{22}}{e_{12}e_{21}} = 1 \qquad (8.5)$$

This product property of expectations for independence may be compared with the additive property of expectations when there is no interaction in the linear model. Indeed, there is exact equivalence of models if we take logarithms of the e_{ij} and write $\tilde{x}_{ij} = \ln e_{ij}$. This is the basis of the so-called **log-linear model**.

Since (8.5) is a necessary condition for independence, it follows that any hypothesis specifying dependence implies a more general relationship

$$\frac{e_{11}e_{22}}{e_{12}e_{21}} = k \qquad (k \neq 1) \qquad (8.6)$$

On taking logarithms we have the analogue of the interaction model.

8.3.3 Extension to a $2 \times 2 \times 2$ table

The model can be extended to $r \times c$ tables and more importantly to multi-way tables. We confine our discussion to $2 \times 2 \times 2$ tables exemplified by the data in Table 8.9.

First, we outline extension of the linear model to three factors each at two levels in an analysis of variance context. The interaction measure introduced in the two-factor model is called a **first-order** (or sometimes a **two-factor**) interaction. If we have three factors each at two levels, we may represent the expected yield in an obvious extension of the notation of Section 8.3.1 by $\tilde{x}_{ijk}$, $i, j, k = 1, 2$.

If we consider the first two factors at the first level of the third factor (indicated by $k = 1$) we will have a first-order interaction between factor 1 and factor 2 at this fixed level of factor 3 if $\tilde{x}_{111} + \tilde{x}_{221} - \tilde{x}_{121} - \tilde{x}_{211} = I$, $I \neq 0$.

If, in addition, we have a first-order interaction between factor 1 and factor 2 at the second level of factor 3 ($k = 2$), this implies $\tilde{x}_{112} + \tilde{x}_{222} - \tilde{x}_{122} - \tilde{x}_{212} = J$, $J \neq 0$. If $I = J$ we say there is no second-order interaction between the three factors. If $I \neq J$ we say there is a **second-order** or **three-factor** interaction. If $I = J = 0$ we have no interaction.

In the context of the log-linear model where we write $\tilde{x}_{ijk} = \ln e_{ijk}$ where e_{ijk} is the expectation for level i of classification 1, for level j of classification 2 and for level k of classification 3, the no-interaction model corresponds to that for independence and in terms of the e_{ijk} becomes

$$(e_{111}e_{221})/(e_{121}e_{211}) = (e_{112}e_{222})/(e_{122}e_{212}) = 1$$

Dependence may be first or second order (referred to as first- or second-order interaction in the log-linear sense). For the first-order interaction model

$$(e_{111}e_{221})/(e_{121}e_{211}) = (e_{112}e_{222})/(e_{122}e_{212}) = k$$

where $k \neq 1$, and for the second-order interaction model

$$(e_{111}e_{221})/(e_{121}e_{211}) \neq (e_{112}e_{222})/(e_{122}e_{212})$$

The log-linear model allows tests to determine if data like those in Table 8.9 can be adequately described by some particular model. We have already noted that the data do not satisfy an independence or no-interaction model. Does a first-order interaction model fit the data? How do we estimate the e_{ijk} under such a model? What test criteria do we use?

The theory needed for complete answers is beyond the scope of this book, as is the extension to more elaborate situations like that giving rise to the data in Table 8.8. The reader should consult a specialist text for details such as Bishop, Fienberg and Holland (1975), Fienberg (1980) or Plackett (1981).

For our illustrative example we summarize the relevant procedure for testing a first-order interaction model. In the case of a $2 \times 2 \times 2$ contingency table the maximum likelihood estimators of e_{ijk}, which we shall denote by $\hat{e}_{ijk}$, must satisfy the condition

$$(\hat{e}_{111}\hat{e}_{221})/(\hat{e}_{121}\hat{e}_{211}) = (\hat{e}_{112}\hat{e}_{222})/(\hat{e}_{122}\hat{e}_{212}) \tag{8.7}$$

with the constraints that they must also sum to the observed marginal totals over any suffix. In general, maximum likelihood estimates can only be obtained by iterative methods (a well-known one is called the **iterative scaling procedure**); however, in the special case of the $2 \times 2 \times 2$ table with a first-order interaction model, direct calculation is possible.

Once the $\hat{e}_{ijk}$ have been calculated, statistics that are obvious extensions of T or T_1 given in Sections 8.2.2 and 8.2.3 for the $r \times c$ table are used for significance tests. T_1 is often preferred because of certain additive properties that enable it to be partitioned into components analogous to the way this is done for orthogonal sums of squares in the analysis of variance. Illustration of this partitioning is again beyond the scope of this book, but we can establish that a first-order interaction model fits Bartlett's data in Table 8.9.

Example 8.3

The problem. Determine whether the data in Table 8.9 are consistent with a first-order interaction model.

Formulation and assumptions. We postulate a first-order interaction model, calculate all $\hat{e}_{ijk}$ to give maximum likelihood estimates satisfying (8.7), then test using the T_1 statistic.

Procedure. Since the expectations must all add up to the relevant marginal totals we can express all $\hat{e}_{ijk}$ in terms of $\hat{e}_{111}$; substitution in (8.7) then leads to a cubic equation which we solve for $\hat{e}_{111}$. It is easily verified by calculating the relevant marginals that

$$\hat{e}_{121} = \hat{e}_{211} = 240 - \hat{e}_{111}, \quad \hat{e}_{221} = \hat{e}_{111}, \qquad \hat{e}_{112} = 263 - \hat{e}_{111},$$

$$\hat{e}_{122} = \hat{e}_{111} - 125, \qquad \hat{e}_{212} = \hat{e}_{111} - 23, \quad \hat{e}_{222} = 365 - \hat{e}_{111}$$

whence, temporarily denoting $\hat{e}_{111}$ by x, (8.7) may be written

$$\frac{x^2}{(240 - x)^2} = \frac{(263 - x)(365 - x)}{(x - 125)(x - 23)}$$

This cubic equation in x may be solved numerically with an appropriate computer algorithm. The relevant solution is $x = \hat{e}_{111} = 161.1$. The remaining expectations may be calculated by the relations given above and are summarized in Table 8.11.

Table 8.11 Expected numbers under first-order interaction model for the data of Table 8.9

Length of cutting	Long		Short	
Time of planting	Early	Late	Early	Late
Alive	161.9	78.9	101.9	36.1
Dead	78.9	161.1	138.1	203.9

Calculating T_1 (where summation is over all 8 cells) we find $T_1 = 2.30$ (rounding to two decimal places). In this example $T = 2.27$, and is close to T_1. Although we do not prove it, each of these statistics under the null hypothesis has approximately a chi-squared distribution with 1 degree of freedom.

Conclusion. Since T_1 and T are well below the chi-squared value of 3.84

required for significance at the 5% level we accept the hypothesis that the data are consistent with a first-order interaction model.

Important theoretical points have been glossed over in this brief presentation of log-linear models, but the topic is worth including since, although now widely used by statisticians working with categorical data, it is less familiar to experimenters who analyse their own data. The methods are applicable to partial contingency tables where large blocks of cells are of necessity zero and where simple models of independence clearly do not hold. A classic example on fingerprint patterns which led to some controversy over 50 years ago is discussed by several authors including Plackett (1981) and in the entry on contingency tables in Kotz and Johnson (1983, vol. 3, pp. 161–71). A log-linear model proves illuminating for these data.

8.4 GOODNESS OF FIT TESTS FOR DISCRETE DATA

In Table 8.2 we listed penultimate digits of telephone numbers, suggesting we may want to test the hypothesis that the digits were random, i.e. that any digit may occupy the position with equal probability.

Tests of goodness of fit to any discrete distribution (often one appropriate to counts) may be made using the chi-squared statistic T in (8.2). The expectations are calculated to accord with the hypothesized distribution which may be the uniform, binomial, negative binomial, Poisson or any other discrete distribution. Sometimes one or more parameters will be estimated from the data. If we have r counts (or cells), the test will have $r - 1$ degrees of freedom if no parameters are estimated; if a parameter such as p for the binomial, or the mean λ for a Poisson distribution, is estimated from the data one further degree of freedom is lost for each such parameter.

Tests of the Kolmogorov–Smirnov type are sometimes applied to discrete data, but generally speaking the test criteria are no longer exact and tests are often inefficient.

We illustrate the chi-squared goodness of fit tests for some simple cases.

Example 8.4

The problem. It is often suggested that people's recorded ages at death are influenced by two factors: firstly, the psychological wish to achieve ages recorded by 'decade', e.g. 70, 80, 90 and that a person near the end of the previous age decade who knows their days are numbered will struggle to reach such an age before dying. If this were so, we might expect ages at death with final digits 0, 1 to be more frequent than higher final digits. A second factor is that sometimes elderly people are unsure about their age, tending to round it; e.g. if they are in the mid-seventies they tend to say they are 75 if anywhere from about 72 or 73 to 77 or 78. Similarly a stated age 80 may correspond to a true age a year or two above or below. If these factors operate we would not expect final digits in recorded ages at death to be uniformly distributed. Table 8.12 gives final digits at age of death for the 117 males recorded at the Badenscallie burial ground (see Section A7.2). Is the hypothesis that any digit is equally likely acceptable?

Table 8.12 Recorded last digit of age at death, Badenscallie males

Digit	0	1	2	3	4	5	6	7	8	9
No. of times obs.	7	11	17	19	9	13	9	11	13	8

Formulation and assumptions. In all, 117 deaths are recorded so the expected number of occurrences of each of the 10 digits if all are equally likely is 11.7. Since the denominator in T is always the expected number, $e = 11.7$, we may calculate T by taking differences between the observed numbers, n_i and 11.7, squaring these, adding and finally dividing the total by 11.7. The differences $n_i - e$ are respectively $-4.7, -0.7, 5.3, 7.3, -2.7, 1.3, -2.7, -0.7, 1.3, -3.7$. Note that these differences sum to zero; this follows from a standard property of the mean. Their sum of squares is 136.1. Division by the expected number, 11.7, gives $T = 136.1/11.7 = 11.63$. The value of chi-squared for significance at the 5% level with $10 - 1 = 9$ degrees of freedom is 19.02.

Conclusion. T is well below the critical value; we do not reject the hypothesis that the digits may be random.

Comments. On the basis of our suppositions in posing the problem we would expect a build-up of digits, around 0, 1, and perhaps again at 5. There is variation in the frequencies of the digits but this is not the build-up we observe. Any build-up is around 2 and 3. As the overall result is not significant we should not read much into this, but it may represent some tendency, especially for more elderly people aged about 82 or 83 or 92 or 93 to feel they have no chance of outliving that decade and to make no special efforts to prolong life; but that is pure speculation. There is certainly no evidence in the data that age may be erroneously reported to the nearest half-decade, e.g., as 75, 80, 85, etc.

8.4.1 Goodness of fit with a parameter estimate

We often have data that might belong to some specific distribution such as a Poisson distribution with an unknown mean. We may use the statistic T to test this (although in the particular case of the Poisson and some other distributions there are alternative parametric tests that are more powerful against a wide range of alternatives).

When using the chi-squared test based on the statistic T, a difficulty arises if the expected numbers in some 'cells' are very small. The usual advice is that such cells be banded together to give an expected number close to 5 (but this is only a rough guide) for the grouped cells. There is a corresponding reduction in degrees of freedom.

Example 8.5

The problem. In a factory with 220 employees the numbers of people who have 0, 1, 2, 3, ... accidents in a given year are recorded.

No. accidents	0	1	2	3	4	5	6 or more
No. of people	181	9	4	10	7	4	5

Are these data consistent with a Poisson distribution?

Formulation and assumptions. The maximum likelihood estimate, $\hat{\lambda}$, of the Poisson parameter, λ, is given by the mean number of accidents per person. Expected numbers, $E(X = r)$, having r accidents are given by $E(X = r) = 220\hat{\lambda}^r \exp(-\hat{\lambda})/r!$; this follows because $\Pr(X = r) = \lambda^r \exp(-\lambda)/r!$ (see Section A1.1).

Procedure. The mean number of accidents per person is calculated by multiplying each number of accidents by the number of people having that number of accidents, adding these products and dividing the resulting sum by 220, giving

$$\hat{\lambda} = (0 \times 181 + 1 \times 9 + 2 \times 4 + 3 \times 10 + 4 \times 7 + 5 \times 4 + 6 \times 5)/220$$
$$= 125/220 = 0.568$$

There is an approximation involved, as we have treated '6 or more' as though it were exactly 6. In practice this has little influence on our result; but it is a limitation of our data that we should recognize. A pocket calculator with an e^x key aids calculation of $E(X = r)$. A simple iterative algorithm may be used noting that $E(X = r + 1) = E(X = r) \times \hat{\lambda}/(r + 1)$.

The expected numbers are:

No. accidents	0	1	2	3	4	5	6 or more
Expected nos	124.7	70.8	20.1	3.8	0.5	0.1	0

Grouping results for 3 or more accidents gives an associated total expected number 4.4. We calculate T using formula (8.3) as

$$T = (181^2/124.7) + (9^2/70.8) + (4^2/20.1) + (26^2/4.4) - 220 = 198.3$$

The degrees of freedom are 2, since we have four cells in the final test and we have estimated one parameter λ. Chi-squared tables show that the critical value for significance at the 0.1% level with 2 degrees of freedom is 13.81.

Conclusion. There is overwhelming evidence that the data are not consistent with a Poisson distribution.

Comments. It has long been established that accident data often do not follow a Poisson distribution; they would do so if accidents occurred entirely at random and all people had an equal probability of experiencing an accident, and having one accident did not alter the probability of that person having a further accident. In practice accident data are usually better described by a model that allows for different people having a different degree of accident proneness, or that allows for the occurrence of one accident to alter a person's susceptibility to further accidents, either reducing the probability because they become more careful, or increasing it because the shock of one accident affects a person's concentration, and they are more likely to have another. These factors may act differently for different people, or for people exposed to different risks. Multiple accidents, in which a number of people are involved in the one incident, will also affect the distribution of numbers of accident victims.

8.4.2 Goodness of fit for grouped data

The chi-squared test is sometimes applied to test goodness of fit of grouped data from a specified continuous distribution, commonly a normal distribution with either specified mean and variance or with these estimated from the data. This procedure is not recommended unless the grouped data are the only data available and the grouping is made on some natural basis. One difficulty with using arbitrary groupings is that one grouping may result in rejection of a hypothesis whereas an alternative grouping may not.

An example of a situation where a test for normality using T given by (8.2) and based on a natural grouping might reasonably be considered is that for numbers of sales of clothing of various sizes. For example, a large retailer might note sales of ready-made trousers with nominal leg lengths (cm) 76, 78, 80, ... The implication is that customers requiring an ideal leg length between 75 and 77 cm will purchase trousers of leg length 76, those requiring leg lengths between 77 and 79 will purchase trousers of leg length 78, and so on. The sizes represent the centre values for each group. We may wish to test whether the sales numbers are consistent with a normal distribution. If no mean and variance are specified these must be estimated from the grouped data and 2 degrees of freedom deducted to allow for this.

Given complete continuous sample data it is better to use tests for goodness of fit of the Kolmogorov type or, if appropriate, Lilliefors or other tests relevant to a particular continuous distribution. If grouping is used and tests based on the T statistic (8.2), parameter estimates, if required, should be based on the group data rather than the original data even if the latter are available. Anomalies may arise if this procedure is not adopted; these are discussed by Kimber (1987).

8.5 FIELDS OF APPLICATION

Tests of independence of categorical data occur widely. Here are a few situations.

Rail transport

A railway operator may be worried whether its image is different for standard-class and for first-class passengers. It may ask samples of each to grade the service as excellent, good, fair or poor, setting up a 4×2 table of response numbers.

Television viewing

A government-sponsored service competes with a commercial service. Samples of men and of women are asked which they prefer. The results may be expressed in a 2×2 table, any difference between the sexes in preference ratings being of interest.

Medicine

A doctor compares two treatments for curing drug addiction. With each withdrawal symptoms are classified as severe, moderate or slight. The resulting 2×3 table of numbers in each category is examined to see whether there is evidence of an association between treatments and severity of withdrawal symptoms. If the investigation were extended to four different addictive drugs the doctor would end up with a $2 \times 3 \times 4$ table and might use a log-linear model to study the nature of an interaction between addictive drugs and the responses to the two treatments.

Sociology

A social worker may be interested in whether blonde or brunette teenage girls who frequent public houses are more likely to drink spirits; a 2×2 table with categories spirits/non-spirits and blonde/brunette may be used to test for lack of association between hair colour and preferred drink.

Public health

After a contaminated food episode on a Jumbo jet where some passengers show mild cholera symptoms, the airline authorities are anxious to know if those previously vaccinated show a higher degree of immunity. They conduct an enquiry to find which passengers have and have not been vaccinated; records also show how many in each category exhibit symptoms, so they can test for association.

Rain making

In low-rainfall areas experiments are carried out in which rain-bearing clouds are 'seeded' to induce precipitation. Randomly selected clouds are either seeded or not seeded and we observe whether or not local rainfall occurs within the next hour.

Educational research

Children are to be exposed to an ordeal which some may fear but which holds no terror for others. It is sometimes argued that an explanation of the nature of the ordeal should be given before it is actually faced. This may reduce fear in some who were initially frightened but implant fear in others who were initially fearless. If numbers who are influenced by an explanation prior to the ordeal can be obtained, McNemar's test may be used to indicate whether an explanation does more harm than good.

Medicine

To compare two drugs, one of which is definitely a placebo and the other a drug that might or might not effect a cure, sufferers might be given the drugs in random order and asked to note which, if either, is effective. Some will claim both are, some neither, some one but not the other. McNemar's test can be used to see if results favour the placebo, the new drug or neither.

Here are examples of goodness of fit tests.

Genetics

Genetic theories often specify the proportion of plants in a certain cross that are expected to have, say, blue or white flowers, round or crinkled seeds, etc. Given a sample in which we know the numbers of each combination of flower colour and seed shape, we may use a chi-squared goodness of fit test to see if these are consistent with theoretical proportions.

Sport

It is often claimed that the starting positions in horse-racing, athletics or rowing events may influence the chance of winning. If we have starting positions and winners for a series of, say, rowing events in which there have always been six starters, we can test the hypothesis that the number of wins from each starting position is consistent with a uniform distribution.

Horticulture

The position at which leaf buds or flowers form on a stem of a plant are called nodes. Some theories suggest a negative binomial distribution for the node number (counted from the bottom of the stem) at which the first flower forms. Data can be used to test this hypothesis by a chi-squared goodness of fit test.

Commerce

A motor salesman may have doubts about the value of advertising. He might compare sales over a considerable period for weeks in which he advertises in 0, 1, 2, 3 or 4 newspapers. He would expect a uniform distribution if advertising were worthless. Note that in this example some allowance should be made for trends with time in sales; the effect of this should be minimized if the weeks were chosen at random over a sufficiently long period.

Queues

Numbers of people entering a post office during one-minute intervals might be recorded over a long period. If the process were completely random with constant mean, the numbers of intervals in which 0, 1, 2, ... people enter should be consistent with a Poisson distribution. In this example would you be surprised if your test rejected the hypothesis of a Poisson distribution? If so, why?

EXERCISES

8.1   In a psychological test for pilot applicants, each person applying is classed as extrovert or introvert and is subjected to a test for pilot aptitude which he or she may either pass or fail. Do the results suggest an association between pilot aptitude and personality type?

	Introvert	Extrovert
Pass	14	34
Fail	31	41

8.2   A manufacturer of washing machines issues instructions for their use in English for the UK and US markets, in French for the French market, German for the German market and Portuguese for the Portuguese and Brazilian markets. The manufacturer conducts a survey of randomly selected customers in each of these markets and asks them to classify the instructions (in the language appropriate to that country) as excellent, reasonable or poor. The responses are set out in Table 8.13. Does the survey indicate the instructions are more acceptable in some countries than others?

Table 8.13 Responses to washing machine instructions

	Excellent	Reasonable	Poor
UK	42	30	28
USA	20	41	19
France	19	29	12
Germany	26	22	12
Portugal	18	31	21
Brazil	31	42	7

(It is sensible to consider UK/USA separately – similarly Portugal/Brazil – because of differences in idiom that may affect understanding of the same instructions, e.g. British cars have 'boots' and run on 'petrol'; American cars have 'trunks' and run on 'gas'.)

8.3   Prior to an England v. Scotland football match 80 English, 75 Scots and 45 Welsh supporters are asked who they think will win the game. The numbers responding each way are:

	English	Scots	Welsh
English win	55	38	26
Scottish win	25	37	19

Is there evidence that the proportions expecting each side to win are influenced by nationality?

8.4   Palpiteria is a country where all who visit a doctor pay a bill. A political party produces the information in Table 8.14 to support its claim that poorer people are inhibited from seeking medical aid because of the cost.

Table 8.14 Visits of wage-earners to doctors

Time since last visit to doctor	Income		
	Over 10 000P	5000–10 000P	Under 50000P
Under 6 months	17	24	42
6–12 months	15	32	45
Over 12 months	27	142	271
Never been	1	12	127

The figures may be taken to relate to a random sample of wage-earners. The incomes are given in Palpiliras (P), the country's unit of currency. Do they substantiate a claim that the poor make proportionately less use of the medical services?

8.5   Would your conclusion in Exercise 8.4 have been different if the data were in only two income groupings: (i) under 5000 P and (ii) over 5000 P, i.e. if columns 1 and 2 had been lumped together in Table 8.14?

8.6   Six boys and six girls are each subjected to the same endurance test. If they endure their suffering for 30 hours without falling asleep they pass; otherwise they fail. Given the results below, use Fisher's exact test to determine whether there is firm evidence of a difference in endurance between the sexes.

	Pass	Fail
Boys	1	5
Girls	4	2

8.7   A machine part is regarded as satisfactory if it operates for 90 days without failure. If it fails in less than 90 days it is unsatisfactory and this results in a costly replacement operation. A supplier claims that the probability of a satisfactory life for each part supplied is 0.95 and all failures are independent. Each machine requires four of these parts and all must be functional for satisfactory operation. In a 90-day test run it may be assumed no part will fail more than once. To test the supplier's claim a buyer runs each of 100 machines initially fitted with four new parts for a 90-day test period. The numbers of parts (0 to 4) surviving 90 days are recorded for each machine with the following results:

No. surviving	0	1	2	3	4
No. of machines	2	2	3	24	69

Do these data substantiate the supplier's claim?

8.8   It is claimed that a typesetter makes random errors at an average rate of 3 per 1000 words set, giving rise to a Poisson process. Random samples of 100 sets of 1000 words from his output are examined and the mistakes in each counted. Are the results below consistent with the above claim?

No. of errors	0	1	2	3	4	5	6	7
No. of samples	6	11	16	18	12	14	12	11

8.9   When personal or highly emotive questions are asked in an interview there is some evidence that answers may be influenced by factors such as the age, sex, social background and race of the interviewer. A random sample of 500 women aged between 30 and 40 are further divided randomly into groups of 100 and each group is allocated to one of the following interviewers:

A. A 25-year-old white female with secretarial qualifications.
B. A middle-aged clergyman.
C. A retired army colonel.
D. A 30-year-old Pakistani lady.
E. A non-white male university student.

Each interviewer asks each of the 100 people allocated to him or her: 'Do you consider marriages between people of different ethnic groups socially desirable?'
The numbers answering 'yes' in each group are given below. Assess the evidence that response might be influenced by the type of person conducting the interview.

Interviewer	A	B	C	D	E
No. of 'yes'	32	41	18	57	36

8.10   To measure abrasive resistance of cloth 100 samples of a fabric are each subjected to a 10-minute test under a series of 5 scourers each of which may or may not produce a hole. The number of holes (0 to 5) is recorded for each sample. Are the data consistent with a binomial distribution with $n = 5$ and p estimated from the data? (Determine the mean number of holes per sample. If this is x, then p is estimated by $x/5$.)

No. of holes	0	1	2	3	4	5
No. of samples	42	36	14	3	4	1

8.11   The numbers of deaths in each four-hour period of the day are recorded for one year at a large hospital. Test the hypothesis that the deaths are spread uniformly throughout the day.

Period	Midnt–4 a.m.	4–8 a.m.	8 a.m.–noon
No. of deaths	89	63	29
Period	noon–4 p.m.	4–8 p.m.	8 p.m.–midnt
No. of deaths	36	39	42

8.12 The McNemar test data in Table 8.7 can be reformulated in a way that makes the Cochran test given in Section 6.5.4 appropriate with $t = 2$. Denoting a success by 1 and a failure by 0, we may classify each of the 108 climbers' outcomes for the first and second climb as either 0 or 1 in a 2×108 table. Show that the Cochran Q statistic is in this case identical with the statistic T given in Section 8.2.5.

8.13 In a time-and-motion study 40 trained workers are randomly divided into four groups A, B, C, D each of 10 people. They each carry out their normal routine production task for a day and the number of items produced by each worker is recorded. Group A work in a room with the temperature controlled at 18 °C, group B at 20 °C, group C at 22 °C and group D at a temperature initially at 22 °C but which is allowed to drop steadily during the day to 18 °C.

Use the data below on numbers of items produced by each worker to carry out a median test to see if there is evidence of an effect of temperature on output.

Group A	17	14	21	17	19	18	16	14	19	21
Group B	14	16	19	18	18	17	21	15	16	18
Group C	13	17	16	11	14	15	17	12	14	13
Group D	21	19	18	20	22	17	19	19	20	19

8.14 Aitchison and Heal (1987) give numbers of OECD countries using significant amounts of only 1, or only 2, or only 3 or 4 fuels in each of the years 1960, 1973, 1983. Are the proportions changing significantly with time?

	Year		
No. of fuels	1960	1973	1983
1	7	10	1
2	13	11	13
3	5	4	9
4	0	0	2

(Think carefully about the interpretation of your finding. The data are in no sense a random sample from any population of countries. Your thoughts should concentrate on trends.)

8.15 Marascuilo and Serlin (1979) report a survey in which a number of women were asked to say whether they considered the statement 'The most important qualities of a husband are determination and ambition' to be true or false. The respondents were again asked the same question at a later date. Numbers making the possible responses were as follows:

First response	Second response	Numbers
True	True	523
True	False	345
False	True	230
False	False	554

Is there evidence that experience significantly alters attitudes of women towards the truth of this statement?

8.16 Jarrett (1979) gives the following data for numbers of coal-mine disasters involving 10 or more deaths between 1851 and 1962.

Day of week	Sun	Mon	Tue	Wed	Thu	Fri	Sat
No.	5	19	34	33	36	35	29

Month	Jan	Feb	Mar	Apr	May	Jun	Jul
No.	14	20	20	13	14	10	18

Month	Aug	Sep	Oct	Nov	Dec
No.	15	11	16	16	24

Test whether accidents appear to be uniformly spread over days of the week and over months of the year. What are the implications of your findings? Do they surprise you?

8.17 O'Muircheartaigh and Sheil (1983) give the following data for numbers of players with scores (i) par or better (P_1), (ii) over par (P_2), for low- and high-handicap golfers under two wind conditions, W_1 and W_2. Are the data adequately described by a first-order interaction model?

	Low handicap		High handicap	
	P_1	P_2	P_1	P_2
W_1	9	35	1	49
W_2	37	51	12	115

8.18 Dansie (1986) gives data for a survey in which 800 people were asked to rank four makes of car A, B, C, D in order of preference. The number of

times each rank combination was specified is given below in brackets after each order. Do the data indicate that preference may be entirely random? Is there a significant preference for any car as first choice?
ABCD (41), ABDC (44), ACBD (37), ACDB (36), ADBC (49), ADCB (41), BACD (38), BADC (38), BCAD (25), BCDA (22), BDAC (33), BDCA (25), CABD (31), CADB (26), CBAD (40), CBDA (33), CDAB (33), CDBA (35), DABC (23), DACB (39), DBAC (30), DBCA (21), DCAB (26), DCBA (34).

8.19 Noether (1987b) asked students to select by a mental process what they regarded as random pairs from the digits 1, 2, 3, repeating that process four times. Noether recorded frequency of occurrences of the last digit pair written down by each of 450 students.
The results were:

		Second digit		
		1	2	3
Second digit	1	31	72	60
	2	57	27	63
	3	53	58	29

What would be the expected numbers in each cell of the above table if digits were truly random? Test whether one should reject the hypothesis that the students are choosing digits at random. How do you interpret your finding?

8.20 Howarth and Curthoys (1987) give numbers of male and female students in English and Scottish universities in the years 1900–01 and 1910–11. Are proportions in the sexes independent between countries at each date? Would you consider a first-order interaction model necessary or sufficient to explain the observations?

	1900–01		1910–11	
	M	F	M	F
England	11 755	2 090	16 038	3 579
Scotland	4 432	719	5 137	1 599

9

Robustness, jackknives and bootstraps

9.1 THE COMPUTER AND ROBUSTNESS

In this chapter we adopt a different approach and review several important techniques with applications ranging from simple location estimates and tests to complicated regression problems. All require appropriate computer programs for realistic applications and a general understanding of fairly advanced statistical theory to make best use of them and to avoid some of the pitfalls that go with statistical sophistication (indeed with sophistication in most applied science). We can, however, indicate the principles by numerically trivial examples that demonstrate the rationale of each approach. The experimenter who is not a statistical expert should be aware these methods exist; he or she may then seek a statistician's advice about their application to particular problems.

9.1.1 A special kind of robustness

In its broadest sense robustness implies our analysis does not depend too critically on specific distributional assumptions. It is clear from considerations like Pitman efficiency that a wrong assumption may make a procedure less efficient than we would hope, and further that some procedures may be robust against certain types of departure from a model but not against others, e.g. a test or estimation procedure may be robust against long-tail departures from assumptions, but sensitive to skewness. During the last two decades interest has centred on procedures that are robust against 'small' departures from distributional assumptions, where 'small' is interpreted to mean either: (i) a small proportion of the observations may be grossly in error; or (ii) an appreciable proportion of the observations are subject to small perturbations.

Outliers provide a typical example of the first situation. There are many tests for outliers, but some of these themselves lack robustness; they may, for example, be notoriously bad at detecting more than one outlier in the same tail; others tend to miss outliers if a pair lie in opposite tails. A simple and reasonably robust test is to classify as an outlier any observation x_0 for which

$$|x_0 - \text{med}\,(x_i)|/\text{med}\,[|x_i - \text{med}\,(x_i)|] > 5 \qquad (9.1)$$

where med(x_i) is the median of all observations. The denominator in (9.1) is a measured of spread called the **median absolute deviation**, often abbreviated to MAD – unfortunate because it is in fact a sensible measure and is robust against outliers. The choice of 5 as a critical value is somewhat arbitrary; motivated by the reasoning that if our observations, apart from the outlier, have an approximately normal distribution it picks up as outliers observations more than about 3 standard deviations from the mean.

Example 9.1

The problem. Use (9.1) to detect any outliers in the data set:

$$8.9 \quad 6.2 \quad 7.2 \quad 5.4 \quad 3.7 \quad 2.8 \quad 17.2 \quad 13.7 \quad 6.9$$

Formulation and assumptions. It is convenient to order the observations, then determine the median and the MAD. We first test the observation furthest from the median using (9.1). If this is an outlier, we test the next observation (in either tail) furthest from the median, proceeding this way. Once we decide an observation is not an outlier, there are no further outliers.

Procedure. The ordered observations are

$$2.8 \quad 3.7 \quad 5.4 \quad 6.2 \quad 6.9 \quad 7.2 \quad 8.9 \quad 13.7 \quad 17.2$$

The median is 6.9 and the absolute deviations from the median are $|2.8 - 6.9| = 4.1$, and similarly 3.2, 1.5, 0.7, 0, 0.3, 2.0, 6.8, 10.3. It is left as an exercise to order these and check that their median, the MAD, is 2.0. Setting $x_0 = 17.2$ in (9.1) the left-hand side becomes $(17.2 - 6.9)/2 = 10.3/2 = 5.15$; our test classes 17.2 as an outlier. It is easily verified that no other observation satisfies the outlier criterion.

Conclusion. The observation 17.2 is an outlier.

Comments. Having decided 17.2 is an outlier our problems are not over. What if all checks show it is a genuine observation? We might conclude this is evidence that our sample is not from a normal distribution, but is the distribution skew, or is it long-tailed symmetric?

When all indications are that an observation is genuine, even if uncharacteristic, there may be a number of reasons for this. Huber (1977) suggests that it is not uncommon for up to 10 per cent of experimental observations to come from a different (and often a longer-tailed) distribution than the bulk of observations, yet both have the same mean. This may occur, for example, if several different people make the observations, or different equipment is used for making some observations, or different sources of raw material are used, or observations are made at appreciably different times (all factors that may introduce differing but often unnoticed sources of variability).

Such mixtures of distributions with the majority of observations having, say, a normal distribution with mean μ and standard deviation σ while a few have a normal distribution with the same mean but standard deviation, say, 3σ, are often unexpected and remain undetected.

Computers have compounded the problem of gross errors going undetected. We spoke in Section 3.2 of the dangers of striking adjacent keys when entering data, for example recording not a correct 3.6, but perhaps an incorrect 34.6. It is right and proper to do visual or statistical checks to detect such errors but when we use large data banks with thousands (or even millions) of data, such mistakes do slip through.

Departures from, say, a normality assumption, due to grouping or rounding may be less severe in effect but cannot be ignored entirely; these are examples of small perturbations affecting most of the data.

Even with basically reasonable models, we are often aware that there may be small departures in the sense of one or two gross departures or a large number of small perturbations, so we want estimation procedures that are almost optimal if we have no such upsets and are little affected if we do have such disturbances.

One such class of estimators are the Huber–Hampel m-estimators, the 'm' because they are like the optimal maximum likelihood estimators when these are appropriate, and are little disturbed by gross departures of a few, or small perturbations in many observations.

9.1.2 Motivation for m-estimators for location

As already indicated, use of m-estimators in any but the simplest problems requires adequate computer programs. We demonstrate the idea behind them by discussing the estimation of the mean of a symmetric distribution. In this case the maximum likelihood estimator is the sample mean; in particular, maximum likelihood estimation of the mean of a normal distribution, given a sample of n observations $x_1, x_2, \ldots, x_n$, is equivalent to least squares estimation where we choose our estimator $\hat{\mu}$ as the value of μ that minimizes:

$$U(x, \mu) = \sum_i (x_i - \mu)^2$$

In other words, $\hat{\mu}$ minimizes the sum of squares of deviations of the observed x_i from it. In mathematical terms $(x - \mu)^2$ is often referred to as a **distance function**, as it provides a measure of the distance of a data point from μ. There are many other possible distance functions, e.g. $|x - \mu|$ expresses distances as absolute distances. While $U(x)$ is minimized by setting $\hat{\mu}$ equal to the sample mean, $V(x, \mu) = \sum_i |x_i - \mu|$ is minimized by taking $\hat{\mu} = \text{med}(x_i)$, the sample median. For samples from a normal distribution we prefer the estimator given by minimizing U, because it has a smaller variance for given n, implying a higher Pitman efficiency.

However, even one outlier has much more effect on the estimator $\hat{\mu}$ derived from U than on that derived from V. Indeed the influence of any x_i upon the estimator $\hat{\mu}$ is linearly dependent on the distance of x_i from μ, i.e. an outlier lying 6 standard deviations from the mean has twice the influence of one lying 3 standard deviations from the mean. We shall not prove this intuitively reasonable result which follows from a study of the **influence function** – an advanced mathematical concept. It is easily illustrated by a simple example.

Example 9.2

Consider the five observations 0, 2, 2, 3, 5. If we estimate μ by the sample mean $\bar{x}$, clearly $\bar{x} = 2.4$. Suppose we made the mistake of recording the data as 0, 2, 2, 3, 15; then $\bar{x} = 4.4$. If we recorded it as 0, 2, 2, 3, 45 (the result of hitting two adjacent keys simultaneously!) we find $\bar{x} = 10.4$. This demonstrates the linear effect; altering an entry from 5 to 15 (i.e. by 10) increases $\bar{x}$ by 2; i.e. from 2.4 to 4.4. If the influence depends linearly on distance we would therefore expect an increase of 3 times 10, i.e. 30 (e.g. from 15 to 45) to increase $\bar{x}$ by 3 times 2, i.e. 6). This is exactly what happens, since $\bar{x}$ increases from 4.4 to 10.4.

On the other hand, if we estimate μ by minimizing $V(x, \mu)$, we choose the sample median which, in each of the three cases, is 2. Indeed, merely pushing one, or indeed more than one, entry further into the tail does not alter our median estimate of μ.

Our efficient estimator of the mean of a symmetric distribution is sensitive to a small number of aberrant observations taking the form of outliers, whereas our less efficient estimator may be unaffected by the odd outlier.

Recognizing this inspired Huber (1972) and others to develop m-estimators that have almost all the optimum properties of the best estimators when there are no data aberrations and retain these (almost) when there are outliers, because they are insensitive to them.

This is achieved in the case of location estimators by choosing an estimator that behaves like the least squares (or equivalent maximum likelihood estimator) for the 'good' observations, but more like the median estimator in its treatment of outliers, in that it reduces their influence by downweighting them. The great advantage of m-estimators is that as well as being robust their efficiency is almost as high as that of maximum likelihood estimators, whereas the median estimator pays for its robustness by loss in efficiency.

9.1.3 Huber's m-estimator

To describe a simple m-estimator we generalize the idea of a distance function. In practice we usually require certain continuity and differentiability properties for a straightforward approach which some distance functions (including $V(x, \mu)$ given in Section 9.1.2) lack. We shall not discuss such niceties, which can usually be surmounted in practice.

The reader familiar with calculus will know that a minimum of $U(x, \mu)$ regarded as a function of μ, if it exists, is obtained by differentiating U with respect to μ and equating the derivative to zero. This leads to the so-called 'normal' or 'estimating' equation

$$\sum_i [-2(x_i - \mu)] = 0 \qquad (9.2)$$

with solution $\hat{\mu} = (\sum x_i)/n = \bar{x}$.

We define a distance function more generally as a function $d(t)$ such that for all t, (i) $d(t) \geqslant 0$, $d(0) = 0$; (ii) $d(t) = d(-t)$; and (iii) the derivative, $\psi(t) = d'(t)$ exists for all t and is a non-decreasing function of t (although this last condition is relaxed for some m-estimators).

For estimation of the mean μ of a symmetric distribution, Huber (1972) proposed a function $d(t)$ where for some fixed $k > 0$,

$$\begin{aligned} d(t) &= \tfrac{1}{2}t^2 & \text{if } |t| \leqslant k \\ &= k|t| - \tfrac{1}{2}k^2 & \text{if } |t| > k \end{aligned}$$

Clearly d is a distance function, with derivatives of the form

$$\begin{aligned} \psi(t) &= -k & \text{if } t < -k \\ &= t & \text{if } |t| \leqslant k \\ &= k & \text{if } t > k \end{aligned}$$

We estimate μ by choosing $\hat{\mu}$ to satisfy the normal equation

$$\sum_i \psi(x_i - \mu) = 0 \qquad (9.3)$$

If we allow $k \to \infty$, then for all x_i, $\psi(x_i - \mu) = x_i - \mu$ and our estimator is the sample mean. For finite k, the form of $d(t)$ shows that for $|x_i - \mu| \leqslant k$ we are minimizing a $d(t)$ equivalent to that in least squares, while if $|x_i - \mu| > k$ we are minimizing a linear function of absolute differences.

We still have to determine k, and having done so, decide how we solve equation (9.3). In general we require an iterative solution and for any chosen k we can proceed by writing (9.3) in the form

$$\sum_i \frac{\psi(x_i - \mu)}{(x_i - \mu)}(x_i - \mu) = 0 \qquad (9.4)$$

Writing $w_i = \psi(x_i - \mu)/(x_i - \mu)$, (9.4) becomes

$$\sum_i w_i(x_i - \mu) = 0$$

which has the solution

$$\hat{\mu} = \sum (w_i x_i) \bigg/ \left(\sum w_i \right) \qquad (9.5)$$

which is a weighted mean of the x_i. However, the weights are functions of $\hat{\mu}$. Thus we must start with an estimate $\hat{\mu}_0$ of $\hat{\mu}$, and with this calculate initial weights w_{i0} and use these in (9.5) to compute a new estimate $\hat{\mu}_1$. This in turn is used to form new weights w_{i1}, say, and the cycle is repeated until convergence, which is usually achieved after a few iterations. We discuss in more detail the choice of k in Section 9.1.4 but it will often suffice to choose k such that the interval (med $(x_i) - k$, med $(x_i) + k$) embraces between about 70% and 90% of the observations. For moderate or large samples, calculation of an m-estimator is tedious without the requisite computer program; however, we illustrate the method for small data sets.

Example 9.3

The problem. Obtain the Huber m-estimators for the three data sets:

(i) 0, 2, 2, 3, 5; (ii) 0, 2, 2, 3, 15; (iii) 0, 2, 2, 3, 45

Formulation and assumptions. The median in all three sets is 2. By choosing $k = 2$ we find the interval (0, 4) covers in all sets 4 of the 5 observations. We use the iterative procedure described above to determine $\hat{\mu}$ for each set.

Procedure. We demonstrate the procedure for data set (i). We choose the sample median, med $(x_i) = 2$ as our initial estimate $\hat{\mu}_0$. Since $\psi(x_i - \hat{\mu}) = x_i - \hat{\mu}$ if $|x_i - \hat{\mu}| < k$ it follows that for x_i satisfying this condition $w_{i0} = 1$. Thus, when $\hat{\mu} = 2$ and $k = 2$, the observations $0, 2, 2, 3$ have weights 1, but for the observation 5, $x_i - \hat{\mu} = 5 - 2 = 3$, which is greater than $k (= 2)$. For this observation $\psi(x_5 - \hat{\mu}) = k = 2$, whence $w_{50} = 2/3$. Thus our first iteration using (9.5) gives the new estimate

$$\hat{\mu} = [1 \times 0 + 1 \times 2 + 1 \times 2 + 1 \times 3 + (2/3)$$
$$\times 5]/[1 + 1 + 1 + 1 + (2/3)] = 2.22$$

We now repeat the cycle with $\hat{\mu} = 2.22$. Now only values between 0.22 and 4.22 have weights 1, so the observations 0 and 5 are downweighted. For 0 the weight is clearly 2/2.22, and for 5 it is now $2/(5 - 2.22) = 2/2.78$. Calculating the weighted mean with these weights now gives a new estimate $\hat{\mu} = 2.29$. A further iteration gives $\hat{\mu} = 2.32$, a value maintained at the next iteration.

We leave detailed calculation for data sets (ii) and (iii) as an exercise for the reader (see Exercise 9.1). However, we point out that for data set (ii) the outlying observation 15 is downweighted drastically. In the first iteration based on the median estimate $\hat{\mu}_0 = 2$, it has weight 2/13 and this changes only slightly in subsequent iterations. The iterations converge to $\hat{\mu} = 2.33$. For the third data set the Huber estimate is again $\hat{\mu} = 2.33$.

Conclusion. The Huber estimates of μ are respectively 2.32, 2.33, 2.33.

Comments. These estimates are virtually the same for each set, whereas we saw in Example 9.2 that the sample means were respectively 2.4, 4.4 and 10.4. In this example the sample median is in each case 2 so this is another location estimator unaffected by an outlier. As indicated in Section 9.1.2 (although we shall not prove it) the median has lower Pitman efficiency than the sample mean, and in general lower efficiency than the Huber estimator when sampling from a normal distribution.

9.1.4 Choice of k and other distance functions

Generally speaking, our $k > 0$ is chosen to downweigh outliers. A function called the influence function confirms (as we might expect) that the smaller we choose k, the less the influence of 'way out' observations. However, if we choose k too small and give full weight to only a few observations near μ, our estimator may be appreciably influenced by rounding or grouping effects in these observations.

There is one further difficulty with the Huber estimator and arbitrary k; it concerns changes in scale. It is well known that if we replace all x_i by $y_i = ax_i + b$ where a, b are constants, then the least squares estimate of the mean for the y_i, say $\hat{\mu}_y$, is given in terms of the estimated mean $\hat{\mu}_x$ for the x_i, by $\hat{\mu}_y = a\hat{\mu}_x + b$. A similar relationship holds for sample medians. This useful property does not hold for m-estimators, but we may recover it by a simple modification of the function ψ, replacing $\psi(x - \mu)$ by $\psi[(x - \mu)/s]$, where s is some measure of spread. Experience has shown an appropriate one to be the median absolute deviation (MAD) divided by 0.6745. This divisor makes s a consistent estimator of σ if the underlying distribution is normal. The weights in (9.5) are modified by use of $(x - \mu)/s$ in the denominator also. With this modification (which can be looked upon as a standardization of the data), choosing $k = 1.5$ gives an estimator with desirable properties for downweighting outliers without overemphasizing the influence of moderate grouping or rounding. Since k is now relevant to values of $(x_i - \hat{\mu})/s$ the latter must be recalculated for each sample value at each iteration. There is no change in s or k between iterations. For the data sets in Example 9.3 (see Exercise 9.2) the estimates of μ using this approach turn out to be 2.34 in each case.

Hampel, in Andrews *et al.* (1972), and others, have suggested alternative distance functions that ignore entirely very extreme observations. Effectively, Huber's distance function limits the influence of extreme values to what they would have if at distance k from $\hat{\mu}$. Hampel's function gives even less influence to extreme values and indeed zero weight to some very extreme ones.

Maximum likelihood estimation is a special case of m-estimators (in the case of estimating μ for the normal distribution it corresponds to setting $k = \infty$ in the Huber estimator).

Andrews (1974) developed m-estimators for regression problems using yet another distance function. Multiple regression is an area where it is notoriously difficult to spot outliers without an elaborate analysis, and m-estimators have a useful role to play, although there may be convergence problems. Solutions are again achieved by what is effectively iterated weighted least squares; obtaining starting values that give reasonable guarantees of convergence is often a large part of the computational effort. Here nonparametric methods like those of Agee and Turner (1979) may prove useful.

For a fuller discussion of m-estimators and influence functions see Randles and Wolfe (1979, Section 7.4), or Huber (1972, 1977) for more mathematical detail of the method and properties.

9.1.5 Trimmed means

A simpler and intuitively appealing method of dealing with outliers or long tails in location problems is by calculating **trimmed means**. The idea is simple. We omit the top $\alpha\%$ and the bottom $\alpha\%$ of the ordered sample values and estimate the population mean using the remaining 'central' observations. Common practice is to trim off the top and bottom 10% (extreme deciles) of the observations, or the first and fourth quartiles; in this latter case the trimmed mean is sometimes called the **interquartile mean**.

There are variants on the trimming procedure. One known as Windsorization shrinks extreme observations inward to the value of the largest remaining observation, thus reducing their influence without ignoring them entirely. Both theoretical and Monte Carlo comparisons have been made between the benefits of trimming and the use of m-estimators. One may conjure up specific examples where one approach fares better than the other but it is probably true by and large if one is dealing with unknown types of data aberration that m-estimators provide a more efficient safety net than trimmed means for location estimates.

9.2 JACKKNIVES AND BOOTSTRAPS

The high speed and relatively low cost of computer calculations has revolutionized attitudes towards obtaining approximate numerical solutions to somewhat intractable mathematical problems. Problems of bias can be tackled this way. **Mean bias** arises when the mean value of the estimator (the value of an estimator of course varies from sample to sample) is not equal to the parameter it is estimating. A simple and well-known example of this is that the maximum likelihood estimator of the variance σ^2 of a normal distribution, i.e. $s^2 = \sum_i (x_i - \bar{x})^2/n$ is biased, having expectation $(n-1)\sigma^2/n$. Bias is removed by the simple expedient of replacing n by $n-1$ in the denominator of s^2. It is perhaps less well known that even after making this adjustment and taking the positive square root, s, to estimate the standard deviation σ, we obtain a biased estimator of σ.

Quenouille (1949) introduced a nonparametric method of estimating bias now known as the **jackknife**. It also provides a method of estimating the variance of a parameter estimate. Closely related to the jackknife, and generally more useful for establishing the variance of parameter estimates, is a procedure called the **bootstrap**. The bootstrap may also be used to obtain confidence intervals for true parameter values.

Both the jackknife and the bootstrap make lavish use of the computer's ability to carry out speedily routine and repetitive calculations. As we did for m-estimators, we content ourselves with an introduction to the rationale of the methods using numerically trivial examples.

9.2.1 The jackknife and bias

Both the jackknife and the bootstrap rely heavily for justification on the fact that the sample cumulative distribution function (see Section A1.4) is the maximum likelihood estimator of a population distribution function $F(x)$. The jackknife is appropriate for estimating bias and variance of estimates that are the sample analogues of the population characteristics they are estimating. We met the sample cumulative distribution function in Sections 3.1.1 and 5.5.1 in connection with Kolmogorov–Smirnov type tests. It is in essence the distribution function for a discrete random variable having a probability mass $p_i = 1/n$ (the size of the step) at each of the ordered x_i. Thus the sample analogue of the population mean is the sample mean $\bar{x} = \sum x_i/n$, and that of the variance is $s^2 = \sum_i (x_i - \bar{x})^2/n$. We have already mentioned that the latter is biased and it is well known that $E(s^2) = \sigma^2 - \sigma^2/n$. The term $-\sigma^2/n$ represents the bias in s^2 as an estimator of σ^2.

Other common sample analogue estimators of population characteristics include, in the bivariate case, the sample correlation coefficient as an estimator of the population correlation coefficient, and the bias of this estimator depends – often in an analytically intractable way – on the particular bivariate population from which we sample. A further estimator commonly met in sampling problems is the estimator $\bar{x}/\bar{y}$ of the ratio of two population means μ_x/μ_y, where again there is bias.

In multiple regression we may be concerned with bias in sample estimates of regression coefficients especially in situations of non-normality, or where we are fitting nonlinear regressions.

So that our discussion may cover any parameter estimate of interest – mean, variance, standard deviation, ratio of means, correlation coefficient, regression coefficient or whatever – we refer to the parameter being estimated as θ; we write $\hat{\theta}$ for the estimator which is the sample analogue of θ, and $\tilde{\theta}$ for the jackknife estimator defined below.

Given a sample of n observations we first calculate $\hat{\theta}$ for all n observations, then repeat this calculation omitting x_1, obtaining an estimate which we denote by $\hat{\theta}_{(1)}$. An operation of this type is carried out n times, omitting at each step one of the remaining observations (or in the case of bivariate data one observation point (x_i, y_i) at a time, with the obvious multivariate analogue). We denote the estimate omitting observation i by $\hat{\theta}_{(i)}$, and the mean of the $\hat{\theta}_{(i)}$ by $\hat{\theta}_{(.)}$; i.e. $\hat{\theta}_{(.)} = \sum_i \hat{\theta}_{(i)}/n$. The jackknife estimator $\tilde{\theta}$ of θ is

$$\tilde{\theta} = n\hat{\theta} - (n-1)\hat{\theta}_{(.)} \tag{9.6}$$

The jackknife estimate of bias is

$$b = (n-1)(\hat{\theta}_{(.)} - \hat{\theta}) \tag{9.7}$$

It is well known that the sample mean is an unbiased estimator of the population mean μ for any distribution, whereas we have pointed out that the sample variance $s^2 = \sum_i(x_i - x)^2/n$ is a biased estimator of the population variance σ^2. We shall content ourselves by showing for a trivial numerical example that the jackknife estimator of the mean is identical with the sample mean and that the jackknife estimator of variance is the usual unbiased estimator with the divisor n in s^2 above replaced by $n-1$. These results can be shown by simple but tedious algebra to be generally true for μ and σ^2.

Example 9.4

Determine jackknife estimators of the mean and variance of the data set 1, 2, 7, 10.

The sample mean is 5 and the sample variance $s^2 = 13.5$. The usual unbiased estimator, using divisor $n-1$, is $s'^2 = 18$.

To calculate the jackknife mean we require the mean of the sample means omitting one point at a time; omitting points in order, the four means are 19/3, 6, 13/3, 10/3. The mean of these four means is $60/12 = 5$, whence (9.6) gives $\theta = 4 \times 5 - 3 \times 5 = 5$. Also (9.7) gives $b = 0$, consistent with the sample mean being unbiased.

For variance we again calculate sample variances omitting one point at a time. The four values we obtain are 98/9, 14, 146/9 and 62/9. The mean of these four values is 12 whence (9.6) gives $\tilde{\theta} = 4 \times 13.5 - 3 \times 12 = 18$, consistent with the unbiased estimator s'^2. The bias estimate given by (9.7) is -4.5, equal to the true bias.

That these results hold generally for the mean and variance estimators was the motivation for jackknife estimation. In mathematical jargon they are said to eliminate bias of order $1/n$. Unfortunately many estimators such as the sample correlation coefficient for the population correlation coefficient exhibit a more complicated form of bias. The jackknife in these cases does not completely eliminate bias; however, it usually results in an appreciable reduction.

Generally speaking, reduction of bias in estimators is a desirable end in itself, but the subject is a complicated one, for sometimes a biased estimator has a smaller mean square error than an unbiased estimator, where the **mean square error** is defined as $E(\hat{\theta} - \theta)^2$, where $\hat{\theta}$ is an estimator of some parameter θ. Clearly this is a measure of how close our estimator is to the true value and it may be broken down into additive components representing variance and bias. Reducing one of the components sometimes increases the other.

9.2.2 Jackknife estimates of variance

It is well known that the sample mean $\bar{x}$ as an estimator of the population mean μ has variance σ^2/n where n is the population variance. Conventionally we estimate this by the unbiased estimator $\sum_i(x_i - \bar{x})^2/[n(n-1)]$. Tukey (1958) proposed a jackknife estimator for the variance of any parameter estimator $\hat{\theta}$ where θ is the sample equivalent of the population parameter being estimated. We write this estimator

$$\hat{V}(\hat{\theta}) = (n-1)\sum_i(\hat{\theta}_i - \hat{\theta}_{(.)})^2/n \tag{9.8}$$

The expression $(\hat{\theta}_i - \hat{\theta}_{(.)})^2$ is of the form $(x_i - \bar{x})^2$ with $\hat{\theta}_i$, $\hat{\theta}_{(.)}$ replacing x_i, $\bar{x}$. Tedious elementary algebra shows that for $\hat{\theta} = \bar{x}$ this reduces to the usual unbiased estimator for the mean. We verify this for some simple numerical data.

Example 9.5

For the data 1, 2, 7, 10 where the mean is the parameter of interest, we showed in Example 9.4 that $s^2 = 18$ was the unbiased estimator of σ^2, whence it follows that the estimated variance of the sample mean is $s^2/n = 18/4 = 4.5$. To calculate the jackknife estimator we note that the four means with one observation deleted are respectively 19/3, 6, 13/3, 10/3 (see Example 9.4), whence substitution in (9.8) gives $V(x) = 4.5$, the usual unbiased estimator.

While this equivalence was the motivation for the jackknife estimator of variance of a sample estimate and the estimator generally gives reasonable approximations for estimators of other parameters, the method has not proved as useful as was originally hoped. Early suggestions that use of standard deviation estimates based on $\sqrt{[\hat{V}(\hat{\theta})]}$ might reasonably be expected to lead to confidence intervals based on the t-distribution with $(n-1)$ degrees of freedom (as it does exactly for samples from a normal distribution) have been shown to be unsatisfactory for many other parameters and other distributions. This may arise for a number of reasons, e.g. the distribution of the estimators may not be symmetric. However, when the exact theory is intractable the jackknife estimate of the variance or standard deviation of an estimate may prove useful.

If we have $n = gh$ observations where g and h are integers, one may extend jackknife procedures to those in which we omit h observations at a time, doing this for each of g groups. Present indications are that by doing so we get results that show little advantage over omitting observations one at a time.

9.2.3 Bootstrap estimates of variance

Although conceptually simpler than the jackknife, bootstrap estimation of the variance of a parameter estimate usually involves more computation. In the jackknife we essentially calculate the full sample estimate of the parameter and n further estimates omitting one data point each time, a total of $n + 1$ estimates.

For bootstrap estimation, even for small n, we may calculate anything from 100 to many thousands of estimates of a parameter, so implementation is very much a job for the computer.

The principle behind the bootstrap is that we estimate our parameter for each of a number of samples obtained by taking samples of n from our original n data values, these samples being random samples with replacement. This means that in the bootstrap samples, some of our original sample values will be repeated, and some will not occur at all. If our original 9 observations were 3, 5, 7, 11, 12, 13, 15, 19, 21 typical **resamplings** with replacement might (after ordering) be 3, 3, 11, 11, 12, 13, 15, 15, 21 or 3, 5, 7, 7, 7, 11, 12, 19, 19. For each of these we calculate an estimate $\theta*$ of the parameter of interest. In a few cases it is possible to calculate the variance or standard deviation of the bootstrap estimator analytically, but in the majority of cases this is not possible; the bootstrap procedure is then applied by Monte Carlo methods wherein we generate m bootstrap samples (in practice m is usually at least 100); for each of these we calculate the value of $\theta*$. Let that for the ith sample be $\theta*_i$. Denoting the mean of all m bootstrap estimators by $\theta*_{(.)}$ the bootstrap estimate of the variance of $\theta*$ is given by

$$V*(\theta*) = \left[\sum_i (\theta*_i - \theta*_{(.)}) \right] \bigg/ (m - 1) \qquad (9.9)$$

When $m \to \infty$ this converges to the exact bootstrap variance of the sample estimate $\hat{\theta}$, where $\hat{\theta}$ is the sample analogue of the population parameter of interest.

In a number of situations where the true variance or standard deviation of an estimator is known, Efron (1982) found that Monte Carlo estimates with values of m in the range 100 to 1000 were often appreciably closer to the true value than was the jackknife estimate given by (9.8).

Not only may we obtain bootstrap estimates of variance of a parameter estimate, but by Monte Carlo studies one may estimate the distribution of the bootstrap estimates, and if the study is sufficiently large to give accurate tail probabilities, these can be used to obtain bootstrap confidence intervals for estimates. Such estimates have often been found to be close to exact confidence intervals when these can be determined directly, so the method shows great promise in general nonparametric situations when precise assumptions about $F(x)$ are virtually impossible. There is a lot of theoretical work still to be done to establish when such methods are likely to be accurate.

Efron (1982, Ch. 10) discusses exact bootstrap confidence intervals for the population median based on samples of n and shows that these closely approximate the intervals we obtained using sign test procedures in Section 2.1.2.

Efron also discusses a number of other applications of the jackknife and bootstrap and related procedures at a more advanced mathematical level than in this book, but in a way that makes the principles clear even to those who may not follow some of the mathematical arguments.

9.2.4 Cross-validation

One technique of this general type is **cross-validation**. The original use of cross-validation was to split samples randomly into two equal portions and to perform a number of alternative possible analyses (e.g. fit regressions with differing numbers of parameters, with and without data transformations, etc., until one obtained a satisfactory fit with the first half-sample). The method of analysis that best suited this first half-sample was then 'cross-validated' by applying it to the remaining data. Generally speaking one would expect it to perform not quite so well in this case for the method was 'tailored' to fit the first half-sample; however, if the fit was reasonable this gave confidence in the model.

With the advent of computers the method has been extended; in particular it is now used freely to detect aberrant points or individual observations in a manner that superficially resembles jackknifing. The data are analysed repeatedly omitting one (or sometimes a group) of points, in turn, and the effect on estimates is studied. Advocates of jackknifing procedures point out that the resemblance is superficial, in that this form of cross-validation is not concerned with estimating parameter bias or variance, but simply with seeing whether some individuals or groups of observations have a large influence on estimates.

9.3 FIELDS OF APPLICATION

Unlike the earlier chapters where we described applications in a number of different disciplines, we indicate here only the types of statistical problem where the methods have met with success. The fields of application where these problems arise are wide ranging, many of them being suggested by applications elsewhere in this book.

Correlation coefficients

The sample correlation coefficient is generally a biased estimate of its population counterpart. Only in the case of the normal and one or two other distributions is the nature of this bias determined theoretically. In most cases little is known about the precision of estimators (as measured by their variance or standard deviation). Efron (1982) studied the correlation coefficient between performance measures for a sample of 15 law students. The sample estimate of the correlation coefficient ρ was $\hat{\rho} = 0.776$ and the jackknife estimate of bias was -0.007. If normality were assumed the bias estimate is -0.011. In a series of experiments each with 200 samples of 14 from a bivariate normal distribution with $\rho = 0.5$, Efron found that the average estimate of standard deviation of jackknife estimates of ρ for all 200 trials was 0.223 with a standard deviation of 0.085, while for a bootstrap with $m = 512$ it was 0.206 with a standard deviation of 0.063. The true value for the standard deviation of the sample estimate of ρ was, in this case, 0.218.

Ratio estimates

Ratio estimates, $\bar{x}/\bar{y}$ of ratios of population means are widely used in sampling. The jackknife provides a convenient way of estimating their (sometimes considerable) bias as well as the variance of these estimators.

Adaptive trimmed means

Adaptive trimming of means is a procedure whereby analyses are carried out with, say 5%, 10%, 20% trimming of observations in each tail. One is interested in seeing which degree of trimming gives an estimator with the smallest variance. In all cases this variance may be estimated by a jackknife or bootstrap. In sampling experiments where the true standard deviation was known, Efron found that the bootstrap with $m = 200$ performed better than the jackknife for samples of 10 or 20 from several different distributions.

EXERCISES

9.1 Complete the computation of the m-estimators in Example 9.3.
9.2 Using the standardization factor based on the MAD with $k = 1.5$ recalculate $\hat{\mu}$ for the data sets in Example 9.3.

9.3 Determine jackknife estimates of the mean and variance based on a sample $-6, 0, 1, 2, 3$. Verify that the sample variance is a biased estimator of σ^2 and determine the bias.

9.4 Smith and Naylor (1987) gave the following data for strengths of 15 cm lengths of glass fibre and suggested that the two shortest may be outliers. Does the test statistic (9.1) confirm this?

$$
\begin{array}{llllllllll}
0.37 & 0.40 & 0.70 & 0.75 & 0.80 & 0.81 & 0.83 & 0.86 & 0.92 & 0.92 \\
0.94 & 0.95 & 0.98 & 1.03 & 1.06 & 1.06 & 1.08 & 1.09 & 1.10 & 1.10 \\
1.13 & 1.14 & 1.15 & 1.17 & 1.20 & 1.20 & 1.21 & 1.22 & 1.25 & 1.28 \\
1.28 & 1.29 & 1.29 & 1.30 & 1.35 & 1.35 & 1.37 & 1.37 & 1.38 & 1.40 \\
1.40 & 1.42 & 1.43 & 1.51 & 1.53 & 1.61 & & & &
\end{array}
$$

9.5 A sample of two gives values 0, 1. The sample mean $\hat{\theta} = \frac{1}{2}$ is the analogue of the population mean and has bootstrap variance 0.125. Obtain 100 bootstrap estimates of the mean and use these to compute $V^*(\theta^*)$. Is your estimate reasonably close to the bootstrap variance?

9.6 Calculate the mean and variance of the data in Exercise 9.4 and also trimmed means trimming top and bottom (i) deciles, and (ii) quartiles. Explain any differences between these location estimates. Which do you prefer?

9.7 A sample of paired observations of X, Y are

$$
\begin{array}{l}
X = 1 \quad 2 \quad 3 \quad 5 \\
Y = 3 \quad 7 \quad 9 \quad 7
\end{array}
$$

Use the jackknife to estimate the bias in the estimator $\bar{x}/\bar{y}$ of the ratio of population means.

9.8 Use the data in Exercise 9.7 and the jackknife to estimate bias in the sample correlation coefficient as estimate of the population coefficient.

10
Looking ahead

10.1 NONPARAMETRIC METHODS IN A WIDER CONTEXT

We have covered basic nonparametric and distribution-free methods; many of them alternatives to standard parametric techniques. Research workers producing quantitative results will be aware of more sophisticated statistical techniques, again many having parametric and nonparametric counterparts. In this chapter we look briefly at some more advanced nonparametric procedures – some recent. This is not a comprehensive review; that would need a further book.

We indicate some important developments by references to published research; again the references to any one topic are not comprehensive. The reader interested in any particular field will usually find access to earlier work in the latest references.

Development of basic applied nonparametric methods was stimulated by requirements in the social sciences, where ranked data or data distorted by outliers, etc., are common, but many more sophisticated methods have been proposed to deal with problems arising in medical and biological research. This applies particularly to clinical studies involving censored data which we discuss in Section 10.3.6.

There has been a growing realization that techniques developed in diverse fields are often related. Such relationships in nonparametric analogues of analysis of variance are summarized by Meddis (1980).

Most of the methods described in this chapter are best used with guidance from a trained statistician. As it is intended only as a guide to the range of available methods, we assume a wider knowledge of statistical terminology than that needed in the rest of this book.

10.2 DEVELOPMENT FROM BASIC TECHNIQUES

Many basic techniques are undergoing continuing refinement or modification designed either to increase their efficiency or to make them applicable to new problems. We illustrate such developments for selected problems.

Sample size

How large a sample is needed to attain a given power when testing some H_0 against a specific alternative? For many parametric tests this may be found reasonably easily, but the less restrictive assumptions in nonparametric tests sometimes make it more difficult to carry out such power determinations. Noether (1987a) discusses criteria to determine the sample sizes needed for tests with given power against hypotheses that differ from H_0 by an amount that is of interest. These are applicable to the sign test, the Wilcoxon signed rank test, the Wilcoxon–Mann–Whitney test and Kendall's τ (for testing $\tau = 0$ against some fixed alternative). In particular, Noether gives the Pitman efficiency of the sign test relative to the Wilcoxon signed rank test for samples from a number of common distributions: broadly, his findings confirm the known result that the sign test has superior efficiency only for symmetric distributions with extremely long tails. One of these, the Cauchy distribution, is often cited as the classic example of a distribution with infinite variance. On the other hand, the sign test has Pitman efficiency of only 1/3 relative to the Wilcoxon test for samples from a uniform distribution.

A test for several types of difference

In Section 6.2.1 we introduced the Kruskal–Wallis test for location differences between k samples (the analogue of the parametric one-way classification analysis of variance). Boos (1986) has developed more comprehensive nonparametric tests for k samples using linear rank statistics that test specifically for location, scale, skewness and kurtosis.

Simple confidence intervals

While the initial emphasis for one- and two-sample location problems was on hypothesis testing, emphasis in recent years has swung towards estimation. Markowski and Hettmansperger (1982) consider computationally simple and reasonably efficient estimates and confidence intervals for the one- and two-sample location problem.

Bivariate means

Hotelling's T^2-test is a well-known extension of the t-test. It is used for testing hypotheses about a bivariate normal mean (μ_x, μ_y) based on sample values (x_i, y_i). Dietz (1982) gives a nonparametric analogue of Hotelling's T^2.

Exponential scores

We have considered only normal scores. These are inappropriate to data on lifetime distributions where ranking by exponential scores would sometimes be more appropriate. A test using these, the Savage test, is discussed by Lehmann (1975, pp. 103–4).

Tests for extreme value distributions

We mentioned in Section 3.1.3 that when we test goodness of fit to specific families of distributions with unknown parameters, tests of the Kolmogorov–

Smirnov type are not the most efficient. The Lilliefors alternative test for a normal distribution was introduced in that section. More recently tests have been developed that are better than the Kolmogorov test for some specific distributions. Chandra, Signpurwalla and Stephens (1981) give tables of critical values for a test of the Kolmogorov type for the extreme value and Weibull distributions, which play a key role in distributions of lifetimes or times to failure.

Goodness of fit

The Kolmogorov–Smirnov tests (Sections 3.1.1 and 5.5.1) apply strictly to continuous distributions; modifications for discrete distributions for Smirnov tests are discussed by Eplett (1982).

Saunders and Laud (1980) present a multivariate Kolmogorov goodness of fit test.

Symmetry

In the 1970s several workers proposed tests for symmetry. Such tests, based on a Cramer–von Mises type of statistic, are reviewed by Koziol (1980). Breth (1982) gives improved confidence bands for a distribution when the centre of symmetry is known or can be estimated.

Multivariate trends

Dietz and Killen (1981) use the concept of Kendall's τ to develop a multivariate test for trend and discuss its application to pharmaceutical data.

Nomination sampling

An ingenious paper by Willemain (1980) tackles the problem of estimating the median of a population when we have not one sample of n independent observations but instead only the largest observed value in each of n independent samples. Willemain calls this **nomination** sampling. This is sometimes the only type of information available in historical studies or when only selected cases are included in an experiment. For example, in testing a new and scarce or expensive drug, each of n consultants may be offered sufficient to treat only one patient and asked to give it to the patient on their list in greatest need. Records may be available for the level of some blood constituent for each treated patient; we may wish to use these to estimate the mean or median level of this constituent that might be expected if the drug were used for all patients suffering from the disease.

Dominance

Another problem of interest is to determine, for two random variables X, Y the value of $\Pr(X < Y)$ (or vice versa). This is discussed for the continuous case by

Halperin, Gilbert and Lachin (1987) and for categorical data by Simonoff, Hochberg and Reiser (1986).

Minimum response levels

Shirley (1977) proposed a nonparametric test to determine the lowest of a number of different dose levels at which there is a detectable treatment response compared with the level for untreated control subjects. An improved version of this test was proposed by Williams (1986), and a version based on a Friedman type ranking procedure applicable to a randomized block design is given by House (1986).

Small proportion of subjects responding

Johnson, Verrill and Moore (1987) present rank tests when treatments do not affect all units in a sample but there is a response (perhaps an increase) for only a proportion of the units. The situation is common in some biochemical contexts, when not all subjects respond to a stimulus.

Equality of marginal distributions

An ingenious application of the Wilcoxon–Mann–Whitney test was proposed by Lam and Longnecker (1983) to test equality of the marginal distributions when sampling from a bivariate population.

Change-over points

Two areas where there have been important applications of simple nonparametric methods in recent years are to determination of change-over points and to angular data.

The change-over point problem is that in which sample observations are ordered in time and after a fixed time τ they have a distribution which differs in location from that prior to time τ. The problem is to determine τ. The analysis may be extended to several changes in location at times τ_1, τ_2, τ_3, etc. The basic problem is discussed by Pettitt (1979, 1981) and for several possible change-over points by Lombard (1987).

Angular distributions

Problems concerning angular distributions are common in the geophysical sciences and meteorology. At a given weather station are daily wind directions at 6 a.m. correlated with those at 12 noon? Are directions of two kinds of geological fault correlated? Do prevailing wind directions show a seasonal shift? Lombard (1986) discusses the change point problem for angular data. Fisher and Lee (1982, 1983) discuss nonparametric methods of angular association, giving an analogue for Kendall's τ appropriate to angular data.

Spatial medians

An ingenious development of nonparametric methods from the univariate to bivariate case was proposed by Brown (1983). He applies angle tests analogous to the sign test for location shifts in a specified direction, or indeed in any direction, to what is termed the **spatial median**. The spatial median is the location measure in two dimensions that minimizes the sum of absolute distances to observations. Brown develops tests based on the sums of sines and cosines for polar coordinates associated with the directions of points from a hypothetical spatial median.

10.3 MORE SOPHISTICATED DEVELOPMENTS

In this section we outline more sophisticated, often recent, developments in experimental design, regression, the measurement of trend, the use of log-linear models, the jackknife and the bootstrap. We consider also important applications in areas not dealt with specifically in earlier chapters, including applications to censored survival data, to discrimination, and to problems involving mixtures of distributions, etc.

10.3.1 Design and analysis of experiments

Parametric analysis of variance and related multiple comparison tests were quickly extended from the simple one-way classification and randomized block design to special treatment structures such as factorial designs and a variety of non-orthogonal experimental designs.

Nonparametric analysis have to some extent kept pace with these developments and now provide powerful tools when the rather restrictive assumptions necessary for normal theory testing and estimation break down. Durbin (1951) extended Friedman type analyses to incomplete block experiments for ranked data, an extension described by Conover (1980, Section 5.9).

While we gave somewhat unsophisticated multiple comparison tests in Sections 6.4.2 and 6.5.3, improved methods have been developed for randomized block designs by Wei (1982). Shirley (1987) discusses multiple comparisons and the analysis of factorial experiments based on the Kruskal–Wallis approach. A key early paper on multiple comparisons involving rankings is that by Rosenthal and Ferguson (1965).

Some writers including Fligner (1985) have taken a closer look at established nonparametric methods. Fligner makes a case for replacing the Kruskal–Wallis test by one based on all possible pairwise Wilcoxon–Mann–Whitney tests. While an analogous parametric procedure using paired sample t-tests has little to commend it, Fligner shows his procedure has good efficiency relative to Kruskal–Wallis, but for reasons indicated in Chapter 6 care is needed with significance levels as the tests are not independent.

Nonparametric analysis of covariance is described by Quade (1982) and by Conover and Iman (1982).

Klotz (1980) discusses the effect of missing categorical data on Cochran's test described in Section 6.5.4.

Sequential methods analogous to two-stage sampling for quality control are introduced by Spurrier and Hewett (1980) for a test of independence based on Kendall's τ. A first sample is taken and if the evidence from it is insufficient to reach a decision we take a second sample and examine an appropriate combination of the data sets. Lehmann (1975, pp. 103–4) draws attention to other sequential tests involving ranks.

Mack and Wolfe (1981) introduced a k-sample test called the **umbrella** test. This tests $H_0 : F_i(x)$ identical for $i = 1, 2, \ldots, k$, against $H_1 : F_1(x) \leqslant F_2(x) \leqslant \cdots \leqslant F_u(x) \geqslant F_{u+1}(x) \geqslant \cdots \geqslant F_k(x)$ with at least one strict inequality. The umbrella value is u, where $1 \leqslant u \leqslant k$. It extends the idea of ordered location alternatives implicit in Jonckheere's test mentioned in Section 6.2.3.

Experimental design problems for censored data are mentioned in Section 10.3.6.

10.3.2 Regression and curve fitting

In Chapter 7, apart from passing references to multiple regression, we confined our discussion of regression to the bivariate case. We may extend Theil-like procedures to multiple regression, but there are formidable practical problems. The 'orthogonalization' approach of Agee and Turner (1979) is the most promising extension of this type.

However, there are several other nonparameteric or semiparametric approaches to regression problems. Use of m-estimators is described by Andrews (1974) and a generalization of the least squares approach is discussed by Brown and Maritz (1982). Their approach is worth outlining as it is easily followed by those familiar with the rudiments of least squares regression. If we denote residuals to a fitted regression $\hat{y} = a + bx$ by $r_i = y_i - a - bx_i$ then it is easily shown that the least squares estimating equations (see Section A.6) can be written

$$\sum_i r_i = 0$$

$$\sum_i x_i r_i = 0$$

Brown and Maritz replace x_i and r_i in these equations by functions $h(x_i)$, $\phi(r_i)$ that retain the ordering of the x_i and r_i to obtain estimating equations

$$\sum_i \phi(r_i) = 0$$

$$\sum_i h(x_i)\phi(r_i) = 0$$

They consider in particular the following choices:

(i) $h(x_i) = \text{sgn}[x_i - \text{med}(x_i)]$, $\phi(r_i) = \text{sgn}[r_i - \text{med}(r_i)]$

(ii) $h(x_i) = \text{rank}(x_i)$, $\phi(r_i) = \text{rank}(r_i) - \frac{1}{2}(n + 1)$

where $\text{sgn}(x) = 0$ if $x = 0$, $= 1$ if $x > 0$, $= -1$ if $x < 0$, as well as use of the untransformed x_i and r_i.

They obtain tests about, and confidence intervals for, the unknown slope β. They extend their method to two regressor variables.

There are a number of other papers on the use of ranks in regression, a key one being Adichie (1967). Median estimates in simple linear regression are discussed by Brown (1980). The use of Theil's method is further discussed by Maritz (1979).

Comprehensive discussion of much of the theory of distribution-free regression is given in Maritz (1981, Chs. 5 and 6).

For the case where treatments are arranged randomly, Oja (1987) discusses permutation tests in multiple regression and the analysis of covariance.

A completely different approach to nonparametric curve fitting in the nonlinear case uses splines, a sound method for fitting smooth curves to data. Problem of oversmoothing are avoided by the use of penalty functions. We referred in Chapter 7 to a major paper by Silverman (1985) discussing both theoretical and practical aspects, and an interesting practical application to data on growth of sunflowers receiving nitrogen at different times by Silverman and Wood (1987). In an earlier paper Silverman (1984) describes a cross-validation method (see Section 9.2.4) for the choice of smoothing parameters in spline regression. A general review of the uses of splines in statistical problems is given in Wegman and Wright (1983).

Motivated by m-estimators, another method of fitting nonlinear curves is described by Härdle and Gasser (1984). A more sophisticated approach using kernels (essentially weights) in nonparametric regression is that of Gasser, Müller and Mammitzsch (1985).

The nonparametric calibration, or inverse regression, problem in which one uses regression to estimate x corresponding to later observed y is the subject of a paper by Knafl, Sacks, Spiegelman and Ylvisaker (1984).

McCullagh (1980) in a very comprehensive paper deals with a number of regression models for ordinal data that introduce many statistical ideas we have not even touched upon in this book.

For binary response data the logistic regression model is widely used. Here y takes the value 1 or 0 (e.g. 1 = presence and 0 = absence, of a disease) and we attempt to describe likely prevalence of the disease in terms of a vector $\mathbf{x}$ of risk factors. Hastie and Tibshirani (1987) apply nonparametric methods to this model, which is discussed in more general terms by Cox (1970), and with emphasis on using the GLIM computer package, by McCullagh and Nelder (1983).

10.3.3 The log-linear model

There are many papers on use of the log-linear model with categorical data, most fairly specialized in their application. For example, Mickey and Elashoff (1985) consider tests of partial association in $2 \times J \times K$ multi-way tables that generalize an estimator proposed by Mantel and Haenszel (1959).

Example 8.3 demonstrated an elementary application of the log-linear method to a test for interaction in multi-way tables. Fitting hierarchical models in this context is dealt with by Havránek (1984), a subject also tackled by Cox and Plackett (1980).

Kotze and Hawkins (1984) consider possible outliers in two-way tables as a source of apparent interaction and propose testing for interaction in all 2×2 sub-tables as a means of detecting these.

10.3.4 Uses of the jackknife and bootstrap

Uses of these techniques in a number of areas are described in journals concerned with statistical applications. An excellent expository article on the jackknife is given in Kotz and Johnson (1983, vol. 4, pp. 280–5).

An update on the best method for bootstrap confidence intervals is given by Efron (1987).

An idea originating from Efron to 'smooth' the bootstrap by modifying the discontinuity inherent in sample comulative distribution functions is examined critically by Silverman and Young (1987).

Heltshe and Forrester (1983) and Smith and van Belle (1984) discuss bootstrap and jackknife techniques based on quadrat sampling for determining the abundance of plant species in a locality.

The jackknife is used by Harrell and Davis (1982) to estimate quantiles.

10.3.5 Trends, runs, tests for randomness

With the exception of the Cox–Stuart test for trend (Section 2.1.6), and relevant work in Chapter 7, we have not dealt with these topics; not so much because the tests are difficult as because they are wide ranging and the appropriate test depends very much on the question of interest. In particular, there is a virtually unlimited range of tests for randomness; often an elusive goal. Logical difficulties with the concept of randomness are discussed by Ayton and Wright (1987), but these do not cause many problems in practice.

Several basic tests for randomness are for run length, e.g. lengths of runs of values above or below the median; or of even or odd digits. Key papers in this area are those by Mood (1940) and Wald and Wolfowitz (1940).

An example of a specialized type of run test is given by O'Brien and Dyck (1985) who were interested in the problem of deciding whether a sequence of Bernouilli trials (experiments with only two outcomes such as success and failure) were independent with the probability of success remaining constant. Tests for trends in proportions are given by Armitage (1955), and a test due to Noether for cyclical trends is described in Bradley (1968, Section 7.6).

10.3.6 Censored data

There are many situations where data are incomplete in the sense that they are **censored.** This often happens in medical studies where interest centres upon time of survival after onset or detection of a certain disease. For practical reasons such studies must often be terminated (right censored) before all patients have died; in some cases patients will be 'lost' to an experiment (perhaps due to leaving the district or even the country) before observations are complete. If one is interested in the time of onset of a tumour this may not be able to be determined precisely. What will be known is the time of detection; onset can only be said to be before that time (left censoring). In extreme cases a tumour may only be detected after death. In animal experiments where animals have been exposed to, say, a known carcinogen, which may or may not produce a tumour and tumours can only be detected after death, some animals that do not die naturally may be sacrificed (i.e. deliberately killed) and examined to see if a tumour is present.

Determination of death rates due to, say, a cancer, is further complicated by competing hazards; e.g. a person with a malignant tumour may die from a cause unconnected with the cancer itself.

All these complexities result in incomplete data much of which have an element of censoring. Assumptions about times of onset, times of survival, cannot always be precise and subjects may only be ranked for such criteria. Then nonparametric methods are virtually the only ones available.

Key papers in this area with a strong reliance on nonparametric methods, especially those based on ranks, are those by Gehen (1965a, b), Peto and Peto (1972) and Prentice (1978). Peto and Peto's work is extended by Harrington and Fleming (1982).

Different rank tests for survival data are compared and their differences elucidated by a video game in a not too technical paper by Lan and Wittes (1985).

Woolson and Lachenbruch (1983) consider analysis of covariance for censored data, and Hanley and Parnes (1983) consider nonparametric estimation of a multivariate distribution with censoring, while earlier Campbell (1981) considered bivariate estimation.

Dinse (1982) and Dinse and Lagakos (1982) both discuss estimation problems for lifetimes and times to failure for incomplete data, while Emerson (1982) and Brookmeyer and Crowley (1982) discuss confidence intervals for the median of censored data.

Efron (1981) and Robinson (1983) apply the bootstrap to censored data. Pettitt (1983, 1985) discusses aspects of regression with censored data.

Other papers dealing with various refinements of analysing censored data include those by Slud, Byar and Green (1984), Michalek and Mihalko (1984), Turnbull and Mitchell (1984), Woolson (1981), Albers and Akritas (1987), Dinse (1986), Sandford (1985), Woolson and Lachenbruch (1980, 1981); this list is indicative rather than exhaustive.

10.3.7 Some miscellaneous applications

It would require an intensive search of both the statistical literature and of journals in fields such as economics, psychology, biology, medicine to produce a complete list of applications of nonparametric methods, but we demonstrate versatility by mentioning a few widely different fields of application. Discrimination and classification problems are of considerable interest; the first deals with allocation of individual units to known classes on the basis of observed measurements or categorizations, while classification is concerned with determining whether groups of individuals can be broken down into subclasses such that measurements or allotted categories are more similar for members of a particular sub-class than they are between members of different sub-classes. Papers dealing with nonparametric methods of discrimination include those by Aitchison and Aitken (1976) and Hall (1981a).

The paper by Hall and Titterington (1984) deals with nonparametric estimation of mixture proportions where a sample consists of observations some of which come from one distribution, some from another. The distributions may differ in location or scale. A special case of this sort of problem is the outlier situation where some of our observations are supposed to come, for example, from a distribution with larger variance than the other, or from one with a different mean. We may sometimes be content to annul the effect of mixtures by using, say, a robust method or a nonparametric method known to be insensitive to outliers. However, there are occasions where we wish to know the proportion of our observations coming from each distribution. If this is something like 50:50 it makes no sense to talk of observations from either distribution as outliers. Further papers on this topic are by Hall (1981b) and by Titterington (1983).

10.4 THE BAYESIAN APPROACH

Readers familiar with inference theory will know that an important school of inference is the Bayesian school, who regard parameters not as fixed constants but as having a distribution (known as the **prior distribution**) which may be modified on the basis of experimental evidence.

With this emphasis on parameter distributions, the term 'nonparametric Bayesian inference' may seem something of a contradiction. However, Bayesian concepts can be used in a distribution-free or nonparametric context. Ferguson (1973) gave an early survey of applications. In a quantal response bio-assay (where subjects either survive or die) one may assign a prior distribution to the effective dose, an approach described by Ramsay (1972) and by Disch (1981).

There is increasing interest in the theory of nonparametric Bayesian inference and one may expect more reports of practical applications in future. Kestemont (1987) suggests a measure for comparing parametric and nonparametric Bayesian inferences.

Appendix

A1 RANDOM VARIABLES

In Chapter 1 we recommended books that give elementary accounts of basic notions. This appendix does not supplant these, but we summarize a few important concepts referred to frequently in the text that are especially relevant to nonparametric methods. Readers who find these summaries too terse are strongly advised to refer to a recommended text (or another of their choice) for greater detail.

A1.1 Discrete and continuous random variables

A discrete random variable, X, takes only a finite (or countably infinite) set of values $x_1, x_2, x_3, \ldots, x_n$. These values are often integers (perhaps zero and a range of positive integers), when we may write $x_r = r$, say. The $\Pr(X = x_i) = p_i$ is the value of the **probability mass function** (p.m.f.) when $X = x_i$ for any i.

The **cumulative distribution function** (c.d.f.) is

$$F(x) = \Pr(X \leqslant x)$$

If $x_i \leqslant x < x_{i+1}$ then $F(x) = \sum_{r=1}^{i} p_r$.

For the binomial distribution with parameters n, p the p.m.f. is

$$\Pr(X = r) = {}^nC_r p^r q^{n-r}, \qquad r = 0, 1, 2, \ldots, n$$

and the c.d.f. is

$$F(x) = \sum_{r=0}^{[x]} \Pr(X = r) = \sum_{r=0}^{[x]} {}^nC_r p^r q^{n-r}, \qquad 0 \leqslant x \leqslant n$$

where $[x]$ is the **integral part** of x, i.e. the largest integer less than or equal to x, e.g. if $x = 3.72$, $[x] = 3$, but note that if $x = -3.72$ then $[x] = -4$. $F(x)$ is a non-decreasing step function with steps of height ${}^nC_r p^r q^{n-r}$ at integral values of r between 0 and n inclusive. $F(0) = 0$ and $F(n) = 1$.

A useful and easily verified recursive formula for calculating successive values of the p.m.f. for the binomial distribution is

$$\Pr(X = r + 1) = \frac{(n-r)p}{(r+1)q} \Pr(X = r)$$

For the Poisson distribution with parameter λ the p.m.f. is

$$\Pr(X = r) = \lambda^r e^{-\lambda}/r!$$

where $r = 0, 1, 2, \ldots$ and $r! = r \times (r - 1) \times (r - 2) \times \cdots \times 2 \times 1$, when $r \geqslant 1$ and $0! = 1$.

The c.d.f.

$$F(x) = \sum_{r=0}^{[x]} \lambda^r e^{-\lambda}/r! \qquad x \geqslant 0$$

is a step function with step of height $\lambda^r e^{-\lambda}/r!$ at $x = r$. $F(0) = 0$ and $F(\infty) = 1$. For the c.m.f. the recursive formula

$$\Pr(X = r + 1) = \frac{\lambda}{r + 1} \Pr(X = r)$$

is easily verified.

A continuous random variable X may take any value in an interval (a, b), finite or infinite. The **probability density function** (p.d.f.), $f(x)$, is zero for x outside (a, b) and in (a, b), $f(x)\delta x$ represents the probability that x takes a value in the arbitrarily small interval $(x, x + \delta x)$. The **cumulative distribution function** (c.d.f.) is a monotonic non-decreasing function

$$F(x) = \Pr(X \leqslant x) = \int_{-\infty}^{x} f(x)\,dx$$

Always $F(-\infty) = 0$ and $F(\infty) = 1$. If a and/or b are finite then also $F(a) = 0$ and $F(b) = 1$; and $F(x) = 0$ if $x < a$ and $F(x) = 1$ if $x \geqslant b$. A distribution is said to be **symmetric** about θ if, for all x, $f(\theta - x) = f(\theta + x)$.

The normal distribution with mean μ and standard deviation σ is symmetrically distributed about μ.

A1.2 Means, medians and quantiles

The mean value, or expectation, of X is for a discrete distribution,

$$E(X) = \sum_{i} p_i x_i$$

For the binomial distribution with parameters n, p, $E(X) = np$. For the Poisson distribution with parameter λ, $E(X) = \lambda$.

For a continuous distribution,

$$E(X) = \int_{-\infty}^{\infty} x f(x)\,dx$$

For $0 < k < 1$ the kth **quantile** of a random variable X is a value x_k that satisfies the inequalities $\Pr(X < x_k) \leqslant k$ and $\Pr(X > x_k) \leqslant 1 - k$.

For discrete distributions if x_k is not unique, a unique kth quantile is obtained by taking the mean of the largest and smallest x_k satisfying the kth quantile condition. If $k = \frac{1}{2}$ the corresponding quantile is the **median**. If $k = r/4$, $r = 1, 2, 3$, the quantile is the rth **quartile**. Similarly, $k = r/10, r = 1, 2, \ldots, 9$ and $k = r/100, r = 1, 2, \ldots, 99$ define the rth deciles and percentiles respectively. The second quartile, the 5th decile and the 50th percentile correspond to the median.

For the binomial distribution with $n = 5, p = 1/3$ the probabilities associated with each X value are:

r	0	1	2	3	4	5
$\Pr(X = r)$	32/243	80/243	80/243	40/243	10/243	1/243

It is easily verified that $\Pr(X < 2) < \frac{1}{2}$ and $\Pr(X > 2) < \frac{1}{2}$. Thus 2 is the unique median. For the binomial distribution with $n = 5$, $p = \frac{1}{2}$ the probabilities associated with each X value are

r	0	1	2	3	4	5
$\Pr(X = r)$	1/32	5/32	10/32	10/32	5/32	1/32

Clearly $X = 2$ and $X = 3$ both satisfy the conditions for a median so we choose $x_m = (2 + 3)/2 = 2.5$ as the unique median. Figure A1 shows the graph of $F(x)$ for this distribution and illustrates clearly why there is no unique median without this averaging convention.

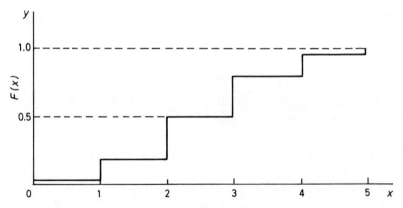

Figure A1 A non-unique median for a binomial variate.

Population means and medians are commonly referred to as **measures of location** or **location parameters.**

A1.3 Higher moments

The kth moment of x about the point a is

$$E\{(X - a)^k\} = \sum_i p_i(x_i - a)^k$$

for a discrete random variable and

$$E\{(X - a)^k\} = \int_{-\infty}^{\infty} (x - a)^k f(x)\,dx$$

for a continuous random variable, where k is a positive integer. This integral may not be convergent for all k. If $a = 0$, $E(X^k)$ is the kth moment about the origin and if $a = \mu$, $E[(X - \mu)^k]$ is the kth moment about the mean and is often denoted by μ_k. The second moment about the mean, μ_2, is the variance. For symmetric distributions the mean is the point of symmetry and for odd k, $\mu_k = 0$. For any distribution $\mu_1 = 0$. A commonly used measure of skewness or asymmetry is $\gamma_1 = \mu_3/\mu_2^{3/2}$. Another feature of interest is the peakedness of distributions relative to the normal; broadly speaking this is measured by $\gamma_2 = (\mu_4/\mu_2^2) - 3$, which has the value zero for the normal distribution. Although not always true, usually, if γ_2 is negative the graph of the probability density function $f(x)$ is flatter than that for the normal distribution, while it is more peaked if γ_2 is positive.

A1.4 The sample cumulative distribution function

Given a random sample of n observations $x_1, x_2, \ldots, x_n$ we may arrange them in ascending order. It is customary to denote the ordered sample values by $x_{(1)}$, $x_{(2)} \ldots, x_{(n)}$ where $x_{(1)} \leqslant x_{(2)} \leqslant \cdots \leqslant x_{(n)}$. The sample cumulative distribution function (sometimes called the empirical c.d.f., and denoted by $S(x)$) provides an estimate of the population c.d.f. and is a step function with step height $1/n$ at each $x_{(i)}$. In the case of r tied values there is a step of height r/n at the tied x value. An example of a sample c.d.f. is given in Figure 3.5.

Sample quantiles are defined with respect to $S(x)$, associating a probability $1/n$ with each sample value. Sprent (1981) gave a modified definition of sample quantiles which has some advantages for small samples, but we here adopt the more widely accepted conventions to obtain uniqueness. Note that if $n = 2m + 1$ is odd, we have a unique sample median at $x_{(m)}$ whereas if $n = 2m$ is even the unique sample median is defined as $\frac{1}{2}\{x_{(m)} + x_{(m+1)}\}$.

A2 PERMUTATIONS AND COMBINATIONS

Many nonparametric tests require a knowledge of the number of possible orderings of a set of n observations, or of all sub-sets of r of these. These are

permutation problems. On other occasions we are not interested in different arrangements within any given sub-set of r, but only want to know how many sub-sets of r (each differing by at least one member) we can form from n observations. This is a **combination** problem.

The number of permutations of r items from n is written ^nP_r and $^nP_r = n(n-1)(n-2)\ldots(n-r+1)$. If we denote four objects $(n=4)$ by A, B, C, D then the possible ordered sets of $r = 2$ are AB, AC, AD, BC, BD, CD together with these pairs in reverse order BA, CA, DA, CB, DB, DC: 12 in all, consistent with $^4P_2 = 4 \times 3 = 12$. The general form for ^nP_r arises because we have n possible first choices, each of which can be combined with one of the remaining $(n-1)$ as second choice to give $n \times (n-1)$ permutations of 2. Continuing in this way we get the general form. When $r = n$, $^nP_n = n!$, the product of all integers from 1 to n.

In a combinations problem we denote the number of groups of r objects from n with at least one different member (where we are not interested in order within each group of r) by ^nC_r, or sometimes by $\binom{n}{r}$. Clearly for each combination of r items there are $^rP_r = r!$ permutations whence $^nC_r \times (r!) = \,^nP_r$, or $^nC_r = (^nP_r)/(r!)$.

It is not difficult to show that

$$^nC_r = n(n-1)(n-2)\ldots(n-r+1)/r! = (n!)/[r!(n-r)!]$$

and that $^nC_r = \,^nC_{n-r}$.

A3 SELECTING A RANDOM SAMPLE OF r ITEMS FROM n

There is no problem selecting a random sample of r items from n if one has a computer that will automatically list a random sample of r numbers between 1 and n inclusive (or between 0 and $(n-1)$). All one has to do is label the n items each with a number between 1 and n (or 0 and $n-1$) as appropriate. The computer program then tells us which numbered items are to form our sample of r. Samples of r may also be got using tables of random digits. If we want a random sample of 12 items from 30 one convenient way to proceed is to number our 30 items to be sampled from 0 to 29. We next select paired random digits from tables. If these lie between 00 (read as 0) and 29 we include the corresponding numbered item in our sample, ignoring any number obtained a second time if we are sampling without replacement. We need not waste digit pairs greater than 29. What we do is divide them by 30 and note the remainder; e.g. 45/30 gives remainder 15, 72/30 gives remainder 12, 60/30 gives remainder zero. These remainders may also be used as sample numbers (again ignoring any already obtained if sampling without replacement). This is fine for digit pairs 30–59 inclusive and 60–89 inclusive, as all remainders between 0 and 29 are equally likely. However, we reject digit pairs 90 to 99 for if we include them we would then have a higher overall probability of remainders 0 to 9 than that for remainders 10 to 29.

Some further dodges enable us to make better use of repeated numbers, but these give little advantage with small samples. If we allow repeated sample values (as when sampling with replacement in bootstrap situations; see Section 9.2.3) we retain repeated values and may twice (sometimes many) times select the corresponding numbered item.

The methods generalize to use of triplets of larger sets of random digits. For example, when we select a sample of 26 items from a total of 123, we may use all triplets from 000 to 983, taking remainders on division by 123, for triplets greater than that number.

A4 The t-DISTRIBUTION AND t-TESTS

A4.1 The single-sample test

If $x_1, x_2, \ldots, x_n$ is a sample of n observations from a normal distribution with unknown mean μ and unknown standard deviation σ, to test $H_0: \mu = \mu_0$ against a one- or two-sided alternative the optimum test statistic is

$$t = \frac{x - \mu_0}{s_x/\sqrt{n}}$$

where $(n-1)s_x^2 = \sum_i x_i^2 - (\sum_i x_i)^2/n$. Under H_0, t has a t-distribution with $n-1$ degrees of freedom. When testing H_0 against $H_1: \mu \neq \mu_0$ large values of $|t|$ indicate significance. Critical values at the 5%, 1% or other levels depend on the degrees of freedom, $n-1$, and are widely tabulated; see e.g. Neave (1981, p. 20). For a one-tail test, t must be taken with the appropriate sign and the critical value appropriate to the required upper or lower tail probability. The distribution of t under H_0 is symmetric, so only positive critical values are tabulated, a change of sign giving the relevant lower tail critical value when that is needed.

A $(100 - \alpha)\%$ confidence interval for μ is given by

$$x \pm t_{n-1,\alpha} s_x/\sqrt{n}$$

where $t_{n-1,\alpha}$ is the critical value of t for significance at the $100\alpha\%$ level in a two-tail test.

A4.2 Two independent samples

If we have two independent samples of size m, n,

$$x_1, x_2, \ldots, x_m$$
$$y_1, y_2, \ldots, y_n$$

from normal distributions with means μ_x, μ_y and the same standard deviation σ then

$$t = \frac{\bar{x} - \bar{y} - \mu_x - \mu_y}{s\sqrt{\{1/m + 1/n\}}}$$

has a t-distribution with $m + n - 2$ degrees of freedom, where $(m + n + 2)s^2 = (m - 1)s_x^2 + (n - 1)s_y^2$, with s_x defined as in Section A4.1 and s_y defined analogously for the second sample. A commonly occurring H_0 is $H_0: \mu_x = \mu_y$, whence

$$t = \frac{\bar{x} - \bar{y}}{s\sqrt{\{1/m + 1/n\}}}$$

Critical values are obtained from t-tables entering these with $m + n - 2$ degrees of freedom.

A $(100 - \alpha)\%$ confidence interval for $\mu_x - \mu_y$ has, in an obvious notation, the form

$$x - y \pm t_{m+n-2,\alpha}s\{\sqrt{(1/m + 1/n)}\}$$

A5 THE CHI-SQUARED AND F-DISTRIBUTIONS

The sum of squares of n independent standard normal variables has a chi-squared distribution with n degrees of freedom.

One reason the distribution is important is because many test statistics in both parametric and nonparametric inference have approximately a chi-squared distribution under the relevant null hypothesis.

Generally speaking, high values of chi-squared indicate significance and critical values relevant to a one-tail test are usually appropriate. These are tabulated at the 5% and 1% level for a wide range of degrees of freedom by Neave (1981, p. 21), who gives examples of how to carry out appropriate tests.

The F-distribution in its simplest form is used to test the hypothesis H_0: $\sigma_x^2 = \sigma_y^2$ on the basis of independent samples of size m, n from normal populations. The test statistic against a two-sided alternative is the larger of $F_1 = s_x^2/s_y^2$ or its reciprocal $F_2 = s_y^2/s_x^2$ where s_x^2, s_y^2 are as defined in Section A4. Under H_0 the relevant statistic has an F-distribution with $m - 1$, $n - 1$ degrees of freedom if s_x^2 is in the numerator, and $n - 1, m - 1$ degrees of freedom if s_y^2 is in the denominator. High values of F_1 or F_2, as appropriate, indicate significance. Critical values for both one- and two-tail tests are given by Neave (1981, pp. 22–5) for a wide range of numerator and denominator degrees of freedom.

In most analysis of variance type situations a one-tail test is appropriate; the numerator in the quotient of estimates of variance is always the one with the higher expected 'population' variance under the alternative H_1. High values using critical levels appropriate to a one-tail test indicate rejection of H_0.

Examples are given by Neave. In most nonparametric tests using F in this book one-tail tests are appropriate, and degrees of freedom for the numerator are given first.

A6 LEAST SQUARES REGRESSION

Given n paired observations (x_1, y_1), $(x_2, y_2), \ldots, (x_n, y_n)$ the straight line of best fit given by least squares is appropriate if we may write our model

$$y_i = \alpha + \beta x_i + \varepsilon_i$$

where the ε_i are unknown 'errors' independently and identically (and usually normally) distributed with mean 0 and (generally unknown) variance σ^2.

The least squares estimates a, b of α, β are obtained by choosing a, b to minimize

$$U = \sum_i r_i^2 = \sum_i (y_i - a - bx_i)^2, \qquad \text{where } r_i = y_i - a - bx_i$$

Equating partial derivatives of U with respect to a and b to zero we get the so-called 'normal' or estimating equation,

$$\sum_i r_i = \sum_i (y_i - a - bx_i) = 0$$

$$\sum_i x_i r_i = \sum_i x_i (y_i - a - bx_i) = 0$$

Solving for a, b we find

$$b = s_{xy}/s_{xx} \quad \text{and} \quad a = \bar{y} - b\bar{x}$$

where

$$s_{xy} = \sum_i x_i y_i - \left(\sum x_i\right)\left(\sum y_i\right)\Big/ n, \qquad s_{xx} = \sum x_i^2 - \left(\sum x_i\right)^2 \Big/ n$$

A7 DATA SETS

We give below two data sets that form the basis of a number of examples and exercises in this book.

A7.1 Numbers of pages in statistics books

Below we record the numbers of pages in each of 114 statistics books on my shelves. A complication that arises is that some books have introductory pages numbered with Roman numerals, i, ii, iii, ... Some number, others do not, a few blank pages at beginning or end. In general the number of pages so involved is small compared to those pages numbered sequentially 1, 2, 3, ... It is the totals of pages so numbered that are recorded below. The numbers of pages are given in the order the books were arranged (at the time of counting) on my shelves.

390	74	189	242	614	296	335	171	427	214	279	429
370	625	287	249	625	783	340	212	686	228	419	467
454	532	283	330	631	709	207	173	230	245	93	433
676	551	246	398	245	511	166	156	593	280	305	284
212	210	307	296	173	153	193	401	240	537	264	402
308	485	365	319	544	168	181	472	361	331	244	354
128	292	181	137	255	312	235	464	503	324	271	312
358	356	110	218	276	425	296	575	248	433	369	142
126	478	157	305	391	412	437	395	258	192	319	197
42	40	34	493	264	199						

A7.2 Age at death of members of four clans

We give below the age at death of male members of four Scottish clans buried in the Badenscallie burial ground in the Coigach district of Wester Ross, Scotland. The data were collected in June 1987. Clan names have been changed but the records are as complete as possible for four real clans. There are inevitably a few missing values as names or dates were unreadable on several headstones, and indeed some headstones appeared to be missing, but these were few in number. Minor spelling variations, especially those of M', Mc, Mac, were ignored. Ages are given for complete years, e.g. 0 means before first birthday, 79 means on or after 79th but before 80th birthday, according to information on the tombstone. Data have been put in ascending order within each clan.

McAlpha (59 names):

0	0	1	2	3	9	14	22	23	29	33	41	41	42	44
52	56	57	58	58	60	62	63	64	65	69	72	72	73	74
74	75	75	75	77	77	78	78	79	79	80	81	81	81	81
82	82	83	84	84	85	86	87	87	88	90	92	93	95	

McBeta (24 names):

0	19	22	30	31	37	55	56	66	66	67	67	68	71	73
75	75	78	79	82	83	83	88	96						

McGamma (21 names):

13	13	22	26	33	33	59	72	72	72	77	78	78	80	81
82	85	85	85	86	88									

McDelta (13 names):

1	11	13	13	16	34	65	68	74	77	83	83	87

A8 TABLES OF CRITICAL VALUES FOR NONPARAMETRIC METHODS

More extensive tables are given by, for example, Neave (1981) and Conover (1980), although the format and actual statistics tabulated may be different. Where quantiles rather than critical values are given these do not usually coincide unless nominal and actual significance levels are the same.

Table A1 Cumulative binomial probabilities, $p = \frac{1}{2}$, $6 \leqslant n \leqslant 20$

$$\Pr(X < r) = \sum_{i=0}^{n} {}^nC_i\left(\tfrac{1}{2}\right)^n \text{ is recorded for given } n, r$$

r \ n	0	1	2	3	4	5	6	7	8	9	10	11	12	13	14	15	16	17	18	19	20
6	0.016	0.109	0.344	0.656	0.891	0.984	1.00														
7	0.008	0.062	0.227	0.500	0.773	0.938	0.992	1.00													
8	0.004	0.035	0.144	0.363	0.637	0.856	0.965	0.996	1.00												
9	0.002	0.020	0.090	0.254	0.500	0.746	0.910	0.980	0.998	1.00											
10	0.001	0.011	0.055	0.172	0.377	0.623	0.828	0.945	0.989	0.999	1.00										
11	0.001	0.006	0.033	0.113	0.274	0.500	0.726	0.887	0.967	0.994	0.999	1.00									
12	0.000	0.003	0.019	0.073	0.194	0.387	0.613	0.806	0.927	0.981	0.997	1.00	1.00								
13	0.000	0.002	0.011	0.046	0.133	0.291	0.500	0.710	0.867	0.954	0.989	0.998	1.00	1.00							
14	0.000	0.001	0.006	0.029	0.090	0.212	0.395	0.605	0.788	0.910	0.971	0.994	0.999	1.00	1.00						
15	0.000	0.000	0.004	0.018	0.059	0.151	0.304	0.500	0.696	0.849	0.941	0.982	0.996	1.00	1.00	1.00					
16	0.000	0.000	0.002	0.011	0.038	0.105	0.227	0.402	0.598	0.773	0.895	0.962	0.989	0.998	1.00	1.00	1.00				
17	0.000	0.000	0.001	0.006	0.024	0.072	0.166	0.314	0.500	0.686	0.834	0.928	0.976	0.994	0.999	1.00	1.00	1.00			
18	0.000	0.000	0.001	0.004	0.015	0.048	0.119	0.240	0.407	0.593	0.760	0.881	0.952	0.985	0.996	0.999	1.00	1.00	1.00		
19	0.000	0.000	0.000	0.002	0.010	0.032	0.084	0.180	0.324	0.500	0.676	0.820	0.916	0.968	0.990	0.998	1.00	1.00	1.00	1.00	
20	0.000	0.000	0.001	0.001	0.006	0.021	0.058	0.132	0.252	0.412	0.588	0.748	0.868	0.942	0.979	0.994	0.999	1.00	1.00	1.00	1.00

Table A2 Critical values for the Wilcoxon signed rank test
Values for the maximum value of the lesser of S_p and S_n for significance at nominal 5% and 1% levels for one- and two-tail tests are given for $6 \leqslant n \leqslant 20$.

n	6	7	8	9	10	11	12	13	14	15	16	17	18	19	20
One-tail test															
5% level	2	3	5	8	10	13	17	21	25	30	35	41	47	53	60
1% level	*	0	1	3	5	7	9	12	15	19	23	27	32	37	43
Two-tail test															
5% level	0	2	3	5	8	10	13	17	21	25	29	34	40	46	52
1% level	*	*	0	1	3	5	7	9	12	15	19	23	27	32	37

*Sample too small for test at this level.

Table A3 Critical values for $|F(x) - S(x)|$ for the Kolmogorov test

Minimum values of $|F(x) - S(x)|$ for significance at the 5% and 1% levels are given for one- and two-tail tests for sample sizes $5 \leqslant n \leqslant 40$ together with an approximation for $n > 40$.

n	5	6	7	8	9	10	11	12	13	14	15	16	17	18	19	20	21	22	23	24	25
One-tail test																					
5% level	0.509	0.468	0.436	0.410	0.388	0.369	0.352	0.338	0.326	0.314	0.304	0.295	0.286	0.279	0.271	0.265	0.259	0.253	0.248	0.242	0.238
1% level	0.627	0.577	0.538	0.507	0.480	0.457	0.437	0.419	0.404	0.390	0.377	0.366	0.355	0.346	0.337	0.329	0.321	0.314	0.307	0.301	0.295
Two-tail test																					
5% level	0.563	0.519	0.483	0.454	0.430	0.409	0.391	0.375	0.361	0.349	0.338	0.327	0.318	0.309	0.301	0.294	0.287	0.281	0.275	0.269	0.264
1% level	0.669	0.617	0.576	0.542	0.513	0.489	0.468	0.449	0.433	0.418	0.404	0.392	0.381	0.371	0.361	0.352	0.344	0.337	0.330	0.323	0.317

n	26	27	28	29	30	31	32	33	34	35	36	37	38	39	40	$n > 40$
One-tail test																
5% level	0.233	0.229	0.225	0.221	0.218	0.214	0.211	0.208	0.205	0.202	0.199	0.197	0.194	0.192	0.189	$1.22/\sqrt{n}$
1% level	0.290	0.284	0.279	0.275	0.270	0.266	0.262	0.258	0.254	0.251	0.247	0.244	0.241	0.238	0.235	$1.52/\sqrt{n}$
Two-tail test																
5% level	0.259	0.254	0.250	0.246	0.242	0.238	0.234	0.231	0.227	0.224	0.221	0.218	0.215	0.213	0.210	$1.36/\sqrt{n}$
1% level	0.311	0.305	0.300	0.295	0.290	0.285	0.281	0.277	0.273	0.269	0.265	0.262	0.258	0.255	0.252	$1.63/\sqrt{n}$

Adapted from a Table by L. H. Miller in *Journal of the American Statistical Association* (1956) by permission of the publisher.

Table A4 Critical values for significance of the Lilliefors test statistic for normality
The values given are the minimum for significance at the 5% and 1% levels in a two-tail test for $6 \leqslant n \leqslant 20$, $n = 25$, 30. Linear interpolation may be used for $20 < n < 30$. An approximation is given for $n > 30$.

n	6	7	8	9	10	11	12	13	14	15	16	17	18	19	20	25	30	$n > 30$
5% level	0.319	0.300	0.285	0.271	0.258	0.249	0.242	0.234	0.227	0.220	0.213	0.206	0.200	0.195	0.190	0.173	0.161	$0.886/\sqrt{n}$
1% level	0.364	0.348	0.331	0.311	0.294	0.284	0.275	0.268	0.261	0.257	0.250	0.245	0.239	0.235	0.231	0.203	0.187	$1.031/\sqrt{n}$

Adapted from a table by H. W. Lilliefors in *Journal of the American Statistical Association* (1967) by permission of the publisher.

Table A5 Critical values for Wilcoxon–Mann–Whitney U statistic (equal sample sizes)
The maximum value of the lesser of U_m, U_n indicating significance in a one- or two-tail test for equal samples sizes $5 \leqslant m = n \leqslant 20$, at nominal 5% and 1% significance levels.

$n = m$	5	6	7	8	9	10	11	12	13	14	15	16	17	18	19	20
One-tail test																
5% level	4	7	11	15	21	27	34	42	51	61	72	83	96	109	123	138
1% level	1	3	6	9	14	19	25	31	39	47	56	66	77	88	101	104
Two-tail test																
5% level	2	5	8	13	17	23	30	37	45	55	64	75	87	99	113	127
1% level	0	2	4	7	11	16	21	27	34	42	51	60	70	81	93	105

Table A6 Critical values for Wilcoxon–Mann–Whitney U statistic (unequal sample sizes)

The maximum value of the lesser of U_m, U_n indicating significance in a one- or two-tail tests for unequal samples sizes between 5 and 20. Values above and to the right of the diagonal apply to significance at a nominal 5% level; values below and to the left of the diagonal apply to significance at a nominal 1% level.

One-tail test

m \ n	5	6	7	8	9	10	11	12	13	14	15	16	17	18	19	20
5		5	6	8	9	11	12	13	15	16	18	19	20	22	23	25
6	2		8	10	12	14	16	17	19	21	23	25	26	28	30	32
7	3	4		13	15	17	19	21	24	26	28	30	33	35	37	39
8	4	6	7		18	20	23	26	28	31	33	36	39	41	44	47
9	5	7	9	11		24	27	30	33	36	39	42	45	48	51	54
10	6	8	11	13	16		31	34	37	41	44	48	51	55	58	62
11	7	9	12	15	18	22		38	42	46	50	54	57	61	65	69
12	8	11	14	17	21	24	28		47	51	55	60	64	68	72	77
13	9	12	16	20	23	27	31	35		56	61	65	70	75	80	84
14	10	13	17	22	26	30	34	38	43		66	71	77	82	87	92
15	11	15	19	24	28	33	37	42	47	51		77	83	88	94	100
16	12	16	21	26	31	36	41	46	51	56	61		89	95	101	107
17	13	18	23	28	33	38	44	49	55	60	66	71		102	109	115
18	14	19	24	30	36	41	47	53	59	65	70	76	82		116	123
19	15	20	26	32	38	44	50	56	63	69	75	82	88	94		130
20	16	22	28	34	40	47	53	60	67	73	80	87	93	100	107	

Two-tail test

m \ n	5	6	7	8	9	10	11	12	13	14	15	16	17	18	19	20
5		3	5	6	7	8	9	11	12	13	14	15	17	18	19	20
6	1		6	8	10	11	13	14	16	17	19	21	22	24	25	27
7	1	3		10	12	14	16	18	20	22	24	26	28	30	32	34
8	2	4	6		15	17	19	22	24	26	29	31	34	36	38	41
9	3	5	7	9		20	23	26	28	31	34	37	39	42	45	48
10	4	6	9	11	13		26	29	33	36	39	42	45	48	52	55
11	5	7	10	13	16	18		33	37	40	44	47	51	55	58	62
12	6	9	12	15	18	21	24		41	45	49	53	57	61	65	69
13	7	10	13	17	20	24	27	31		50	54	59	63	67	72	76
14	7	11	15	18	22	26	30	34	38		59	64	69	74	78	83
15	8	12	16	20	24	29	33	37	42	46		70	75	80	85	90
16	9	13	18	22	27	31	36	41	45	50	55		81	86	92	98
17	10	15	19	24	29	34	39	44	49	54	60	65		93	99	105
18	11	16	21	26	31	37	42	47	53	58	64	70	75		106	112
19	12	17	22	28	33	39	45	51	57	63	69	74	81	87		119
20	13	18	24	30	36	42	48	54	60	67	73	79	86	92	99	

Table A7 Critical values for significance for the Smirnov test (equal sample sizes)

Minimum values of $|S(x) - S(y)|$ for significance in one- and two-tail tests at nominal 5% and 1% for equal sample sizes $5 \leqslant m = n \leqslant 20$. Here and in Table A8 discontinuities often result in appreciable differences between nominal and actual significance levels; this explains apparent discontinuities in the general downward trend of critical values as sample sizes increase.

$m = n$	5	6	7	8	9	10	11	12	13	14	15	16	17	18	19	20
One-tail test																
5% level	0.800	0.833	0.714	0.625	0.667	0.600	0.545	0.500	0.538	0.500	0.467	0.438	0.471	0.444	0.421	0.400
1% level	1.00	1.00	0.857	0.750	0.778	0.700	0.727	0.667	0.615	0.571	0.600	0.563	0.529	0.555	0.526	0.500
Two-tail test																
5% level	1.00	0.833	0.857	0.750	0.667	0.700	0.636	0.583	0.538	0.571	0.533	0.500	0.471	0.500	0.474	0.450
1% level	1.00	1.00	0.857	0.875	0.778	0.800	0.727	0.667	0.692	0.643	0.600	0.625	0.583	0.555	0.526	0.550

Table A8 Critical values for significance for the Smirnov test (unequal sample sizes)

Minimum values of $|S(x) - S(y)|$ for significance in one- and two-tail tests for unequal sample sizes between 5 and 20. Entries above and to right of diagonals are for testing at 5% level; entries below and to the left are for testing at 1% level. See also text for Table A7.

One-tail test

n / m	5	6	7	8	9	10	11	12	13	14	15	16	17	18	19	20
5		0.800	0.714	0.675	0.667	0.700	0.636	0.600	0.615	0.600	0.667	0.600	0.588	0.578	0.589	0.600
6	1.00		0.667	0.625	0.611	0.600	0.576	0.667	0.590	0.571	0.567	0.563	0.649	0.511	0.561	0.550
7	0.857	0.833		0.607	0.571	0.571	0.571	0.547	0.549	0.571	0.533	0.526	0.513	0.516	0.519	0.514
8	0.875	0.833	0.750		0.556	0.550	0.545	0.500	0.519	0.517	0.500	0.563	0.500	0.500	0.487	0.500
9	0.800	0.778	0.746	0.750		0.556	0.525	0.528	0.504	0.500	0.511	0.479	0.484	0.500	0.468	0.467
10	0.800	0.733	0.714	0.700	0.678		0.518	0.500	0.492	0.486	0.500	0.475	0.465	0.456	0.447	0.500
11	0.800	0.742	0.714	0.693	0.636	0.627		0.485	0.469	0.474	0.461	0.455	0.455	0.444	0.440	0.436
12	0.800	0.750	0.690	0.667	0.639	0.617	0.583		0.455	0.464	0.467	0.458	0.441	0.444	0.434	0.433
13	0.769	0.692	0.692	0.644	0.624	0.600	0.601	0.590		0.428	0.446	0.438	0.434	0.423	0.421	0.415
14	0.729	0.714	0.714	0.643	0.635	0.600	0.584	0.560	0.560		0.438	0.428	0.420	0.413	0.414	0.407
15	0.800	0.700	0.667	0.625	0.622	0.600	0.576	0.567	0.574	0.529		0.421	0.412	0.411	0.400	0.417
16	0.738	0.688	0.652	0.688	0.604	0.588	0.568	0.562	0.538	0.536	0.500		0.401	0.403	0.395	0.400
17	0.741	0.667	0.647	0.625	0.601	0.582	0.556	0.549	0.534	0.525	0.514	0.511		0.386	0.390	0.383
18	0.722	0.722	0.659	0.611	0.611	0.578	0.545	0.556	0.526	0.519	0.511	0.493	0.490		0.389	0.378
19	0.737	0.675	0.647	0.612	0.579	0.547	0.545	0.535	0.526	0.508	0.498	0.497	0.489	0.468		0.379
20	0.750	0.667	0.650	0.625	0.578	0.600	0.536	0.533	0.519	0.507	0.500	0.488	0.479	0.472	0.450	

Two-tail test

n \ m	5	6	7	8	9	10	11	12	13	14	15	16	17	18	19	20
5		0.800	0.800	0.750	0.778	0.800	0.709	0.717	0.692	0.657	0.733	0.800	0.647	0.667	0.642	0.650
6	1.00		0.714	0.708	0.722	0.667	0.652	0.667	0.667	0.643	0.533	0.625	0.667	0.667	0.614	0.600
7	1.00	0.857		0.714	0.667	0.657	0.623	0.631	0.615	0.643	0.590	0.571	0.571	0.571	0.571	0.564
8	0.875	0.833	0.857		0.639	0.600	0.602	0.625	0.596	0.571	0.558	0.625	0.566	0.611	0.539	0.550
9	0.889	0.883	0.778	0.764		0.589	0.596	0.583	0.556	0.556	0.556	0.542	0.536	0.556	0.520	0.517
10	0.900	0.800	0.757	0.750	0.700		0.545	0.550	0.569	0.529	0.533	0.525	0.524	0.511	0.495	0.550
11	0.818	0.818	0.766	0.727	0.707	0.700		0.545	0.524	0.532	0.509	0.506	0.497	0.490	0.488	0.486
12	0.833	0.833	0.714	0.708	0.694	0.667	0.652		0.519	0.512	0.517	0.500	0.490	0.500	0.474	0.483
13	0.800	0.769	0.714	0.692	0.667	0.646	0.636	0.608		0.489	0.492	0.486	0.475	0.470	0.462	0.462
14	0.800	0.762	0.786	0.679	0.667	0.643	0.623	0.619	0.571		0.467	0.473	0.467	0.460	0.455	0.450
15	0.800	0.767	0.714	0.675	0.667	0.667	0.618	0.600	0.590	0.586		0.475	0.455	0.456	0.446	0.450
16	0.800	0.750	0.688	0.688	0.653	0.625	0.602	0.604	0.582	0.563	0.554		0.456	0.444	0.437	0.437
17	0.800	0.716	0.706	0.647	0.647	0.624	0.588	0.583	0.576	0.563	0.557	0.526		0.435	0.437	0.429
18	0.778	0.778	0.690	0.653	0.667	0.600	0.596	0.583	0.585	0.556	0.544	0.535	0.536		0.415	0.422
19	0.747	0.728	0.684	0.645	0.626	0.595	0.584	0.570	0.559	0.556	0.533	0.526	0.514	0.515		0.421
20	0.800	0.733	0.664	0.650	0.617	0.650	0.577	0.583	0.550	0.543	0.533	0.525	0.515	0.506	0.492	

Table A9 Critical values for the Spearman rank correlation coefficient

Values given are the minimum values of $|\rho|$ for $5 \leqslant n \leqslant 20$ for significance at the nominal 5% and 1% levels in one- or two-tail tests. An approximation is given which is of moderate accuracy for $21 \leqslant n \leqslant 30$ and good for $n > 30$.

n	5	6	7	8	9	10	11	12	13	14	15	16	17	18	19	20	$n > 20$
One-tail test																	
5% level	0.900	0.829	0.714	0.643	0.600	0.564	0.536	0.503	0.484	0.464	0.446	0.429	0.414	0.401	0.391	0.380	$1.64/\sqrt{(n-1)}$
1% level	1.00	0.943	0.893	0.833	0.783	0.746	0.709	0.678	0.648	0.626	0.604	0.585	0.566	0.550	0.535	0.522	$2.33/\sqrt{(n-1)}$
Two-tail test																	
5% level	1.00	0.886	0.786	0.714	0.700	0.648	0.618	0.587	0.560	0.538	0.521	0.503	0.488	0.474	0.460	0.447	$1.96/\sqrt{(n-1)}$
1% level	*	1.00	0.929	0.881	0.833	0.794	0.764	0.734	0.703	0.679	0.657	0.635	0.618	0.600	0.584	0.570	$2.58/\sqrt{(n-1)}$

*Sample too small for test at this level.

Table A10 Critical values for testing Kendall's tau

Values are given for the minimum of $|n_c - n_d|$ for significance at nominal 5% and 1% levels for one- and two-tail tests, $5 \leqslant n \leqslant 20$.

n	5	6	7	8	9	10	11	12	13	14	15	16	17	18	19	20
One-tail test																
5% level	8	11	13	16	18	21	23	26	28	33	35	38	42	45	49	52
1% level	10	13	17	20	24	27	31	36	40	43	49	52	58	63	67	72
Two-tail test																
5% level	10	13	15	18	20	23	27	30	34	37	41	46	50	53	57	62
1% level	*	15	19	22	26	29	33	38	44	47	53	58	64	69	75	80

* Sample too small for test at this level.

References

Adichie, J. N. (1967) Estimates of regression parameters based on rank tests. *Ann. Math. Statist.*, **38**, 894–904.

Agee, W. S. and Turner, R. H. (1979) Application of robust regression to trajectory data reduction, in *Robustness in Statistics* (eds R. L. Learner and G. N. Wilkinson), London: Academic Press.

Aitchison, J. and Aitken, C. G. G. (1976) Multivariate binary discrimination by the Kernel method. *Biometrika*, **63**, 413–20.

Aitchison, J. W. and Heal, D. W. (1987) World patterns of fuel consumption; towards diversity and a low cost energy future. *Geography*, **72**, 235–9.

Albers, W. and Akritas, M. G. (1987) Combined rank tests for the two sample problem with randomly censored data. *J. Amer. Statist. Assoc.*, **82**, 648–55.

Andrews, D. F. (1974) A robust method for multiple linear regression. *Technometrics*, **16**, 523–31.

Andrews, D. F., Bickel, P. J., Hampel, F. R., Huber, P. S., Rogers, W. H. and Tukey, J. W. (1972) *Robust Estimates of Location. Survey and Advances*, Princeton: Princeton University Press.

Arbuthnot, J. (1710) An argument for Divine Providence, taken from the constant regularity observ'd in the births of both sexes. *Phil. Trans. Roy. Soc.*, **27**, 186–90.

Armitage, P. (1955) Tests for linear trends in proportions and frequencies. *Biometrics*, **11**, 375–86.

Atkinson, A. C. (1986) *Plots, Transformations and Regression: An Introduction to Graphical Methods of Diagnostic Regression Analysis*, Oxford: Clarendon Press.

Ayton, P. and Wright, G. (1987) Tests for randomness? *Teaching Mathematics and its Applications*, **6**, 83–7.

Bahadur, R. R. (1967) Rates of convergence of estimates and test statistics. *Ann. Math. Statist.*, **38**, 303–24.

Bardsley, P. and Chambers, R. L. (1984) Multipurpose estimation from unbalanced samples. *Applied Statistics*, **33**, 290–9.

Bartlett, M. S. (1935) Contingency table interactions. *J. Roy. Statist. Soc., Suppt.*, **2**, 248–52.

Berry, D. A. (1987) Logarithmic transformations in ANOVA. *Biometrics*, **43**, 439–56.

Biggins, J. D., Loynes, R. M. and Walker, A. N. (1987) Combining examination results. *Brit. J. Math. and Statist. Psychol.*, **39**, 150–67.

Bishop, Y. M. M., Fienberg, S. E. and Holland, P. (1975) *Discrete Multivariate Analysis: Theory and Practice*. Cambridge, Mass.: MIT Press.

Boos, D. D. (1986) Comparing K populations with linear rank statistics. *J. Amer. Statist. Assoc.*, **81**, 1018–25.

Bowman, K. O. and Shenton, I. R. (1975) Omnibus test contours for departures from normality based on $\sqrt{b_1}$ and b_2. *Biometrika*, **62**, 243–50.

Bradley, J. V. (1968) *Distribution-Free Statistical Tests*, New Jersey: Prentice-Hall Inc.

Breth, M. (1982) Nonparametric estimation for a symmetric distribution. *Biometrika*, **69**, 625–34.

Brookmeyer, R. and Crowley, J. (1982) A confidence interval for the median survival time. *Biometrics*, **38**, 29–41.

Brown, B. M. (1980) Median estimation in simple linear regression. *Austral. J. Statist.*, **22**, 154–66.

Brown, B. M. (1983) Statistical use of the spatial median. *J. Roy. Statist. Soc.*, B, **45**, 25–30.

Brown, B. M. and Maritz, J. S. (1982) Distribution-free methods in regression. *Austral. J. Statist.*, **24**, 318–31.

Campbell, G. (1981) Nonparametric bivariate estimation with randomly censored data. *Biometrika*, **68**, 417–22.

Carter, E. M. and Hubert, J. J. (1985) Analysis of parallel line assays with multivariate responses. *Biometrics*, **41**, 703–10.

Chandra, M., Singpurwalla, N. D. and Stephens, M. A. (1981) Kolmogorov statistics for tests of fit for the extreme value and Weibull distributions. *J. Amer. Statist. Assoc.*, **76**, 729–31.

Chatfield, C. (1983) *Statistics for Technology* (3rd edn), London: Chapman & Hall.

Cochran, W. G. (1950) The comparison of percentages in matched samples. *Biometrika*, **37**, 256–66.

Cohen, A. (1983) Seasonal daily effect on the number of births in Israel. *Applied Statistics*, **32**, 228–35.

Conover, W. J. (1980) *Practical Nonparametric Statistics* (2nd edn), New York: John Wiley & Sons.

Conover, W. J. and Iman, R. L. (1982) Analysis of covariance using the rank transformation. *Biometrics*, **38**, 715–24.

Cook, R. D. and Weisberg, S. (1982) *Residuals and Influence in Regression*, London: Chapman & Hall.

Cox, D. R. (1970) *The Analysis of Binary Data*, London: Chapman & Hall.

Cox, D. R. and Stuart, A. (1955) Some quick tests for trend in location and dispersion. *Biometrika*, **42**, 80–95.

Cox, M. A. A. and Plackett, R. L. (1980) Small samples in contingency tables. *Biometrika*, **67**, 1–13.

Daniel, W. W. (1978) *Applied Nonparametric Statistics*, Boston: Houghton Mifflin Company.

Dansie, B. R. (1986) Normal order statistics as permutation probability models. *Applied Statistics*, **35**, 269–75.

Dietz, E. J. (1982) Bivariate nonparametric tests for the one sample location problem. *J. Amer. Statist. Assoc.*, **77**, 163–9.

Dietz, E. J. and Killen, T. J. (1981) A nonparametric multivariate test for trend with pharmaceutical applications. *J. Amer. Statist. Assoc.*, **76**, 169–74.

Dinse, G. E. (1982) Nonparametric estimation for partially-complete time and type of failure data. *Biometrics*, **38**, 417–31.

Dinse, G. E. (1986) Nonparametric prevalence and mortality estimators for animal experiments with incomplete cause-of-death data. *J. Amer. Statist. Assoc.*, **81**, 328–36.

Dinse, G. E. and Lagakos, S. W. (1982) Nonparametric estimation of lifetime and disease onset distribution from incomplete observations. *Biometrics*, **38**, 921–32.

Disch, D. (1981) Bayesian nonparametric inference for effective doses in a quantal-response experiment. *Biometrics*, **37**, 713–22.

Durbin, J. (1951) Incomplete blocks in ranking experiments. *Brit. J. Psychol. (Statistical Section)*, **4**, 85–90.

Durbin, J. (1987) Statistics and statistical science. *J. Roy. Statist. Soc.*, A, **150**, 177–91.

Efron, B. (1981) Censored data and the Bootstrap. *J. Amer. Statist. Assoc.*, **76**, 312–19.

Efron, B. (1982) *The Jackknife, the Bootstrap and other Resampling Plans*, Philadelphia: Society for Industrial and Applied Mathematics.

Efron, B. (1987) Better Bootstrap confidence intervals. *J. Amer. Statist. Assoc.*, **82**, 171–85.

Emerson, J. D. (1982) Nonparametric confidence intervals for the median in the presence of right censoring. *Biometrics*, **38**, 17–27.

Eplett, W. J. R. (1982) The distribution of Smirnov type two-sample rank tests for discontinuous distribution functions. *J. Roy. Statist. Soc.*, B, **44**, 361–9.

Ferguson, T. S. (1973) A Bayesian analysis of some nonparametric problems. *Annals of Statistics*, **1**, 209–30.

Fienberg, S. E. (1980) *The Analysis of Cross Classified Categorical Data* (2nd edn), Cambridge, Mass.: MIT Press.

Fisher, N. I. and Lee, A. J. (1982) Nonparametric measures of angular–angular association. *Biometrika*, **69**, 315–21.

Fisher, N. I. and Lee, A. J. (1983) A correlation coefficient for circular data. *Biometrika*, **70**, 327–32.

Fisher, R. A. (1922) On the interpretation of chi-square from contingency tables, and the calculation of *P*. *J. Roy. Statist. Soc.*, **85**, 87–94.

Fisher, R. A. (1935) *The Design of Experiments*, Edinburgh: Oliver & Boyd.

Fisher, R. A. and Yates, F. (1957) *Statistical Tables for Biological, Agricultural and Medical Research* (5th edn), Edinburgh: Oliver & Boyd.

Fligner, M. A. (1985) Pairwise versus joint ranking. Another look at the Kruskal–Wallis statistic. *Biometrika*, **72**, 705–9.

Friedman, M. (1937) The use of ranks to avoid the assumptions of normality implicit in the analysis of variance. *J. Amer. Statist. Assoc.*, **32**, 675–701.

Gasser, T., Müller, H. G. and Mammitzsch, V. (1985) Kernels for nonparametric curve estimation. *J. Roy. Statist. Soc.*, B, **47**, 238–52.

Gat, J. R. and Nissenbaum, A. (1976) Limnology and ecology of the Dead Sea. *Nat. Geog. Soc. Res. Reports – 1976 Projects*, 413–18.

Geffen, G., Bradshaw, J. L. and Nettleton, N. C. (1973) Attention and hemispheric differences in reaction time during simultaneous audio-visual tasks. *Quarterly Journal of Experimental Psychology*, **25**, 404–12.

Gehen, E. A. (1965a) A generalized Wilcoxon test for comparing arbitrarily singly censored samples. *Biometrika*, **52**, 203–23.

Gehen, E. A. (1965b) A generalized two-sample Wilcoxon test for doubly censored data. *Biometrika*, **52**, 650–3.

Hall, P. (1981a) On nonparametric binary discrimination. *Biometrika*, **68**, 287–94.

Hall, P. (1981b) On the nonparametric estimation of mixture proportions. *J. Roy. Statist. Soc.*, B, **43**, 147–56.

Hall, P. and Titterington, D. M. (1984) Efficient nonparametric estimation of mixture proportions. *J. Roy. Statist. Soc.*, B, **46**, 465–73.

Halperin, M., Gilbert, P. R. and Lachin, J. M. (1987) Distribution free confidence intervals for $\Pr(X_1 < X_2)$. *Biometrics*, **43**, 71–80.

Hanley, J. A. and Parnes, M. N. (1983) Nonparametric estimation of a multivariate distribution in the presence of censoring. *Biometrics*, **39**, 129–39.

Härdle, W. and Gasser, T. (1984) Robust nonparametric function fitting. *J. Roy. Statist. Soc.*, B, **46**, 42–51.

Harrell, F. E. and Davis, C. E. (1982) A new distribution-free quantile estimator. *Biometrika*, **69**, 635–40.

Harrington, D. P. and Fleming, T. R. (1982) A class of rank test procedures for censored survival data. *Biometrika*, **69**, 553–66.

Hastie, T. and Tibshirani, R. (1987) Nonparametric logistic and proportional odds regression. *Applied Statistics*, **36**, 260–76.

Havránek, T. (1984) A procedure for model search in multidimensional contingency tables. *Biometrics*, **40**, 95–100.

Heltshe, J. F. and Forrester, N. E. (1983) Estimating species richness using the Jackknife procedure. *Biometrics*, **39**, 1–11.

Hill, I. D. (1974) Association football and statistical inference. *Applied Statistics*, **23**, 203–8.

Hill, N.S. and Padmanabhan, A. L. (1984) Robust comparison of two regression lines and biomedical applications. *Biometrics*, **40**, 985–94.

Hollander, M. and Wolfe, D. A. (1973) *Nonparametric Statistical Methods*, New York: John Wiley & Sons.

Hora, S. C. and Conover, W. J. (1984) The *F*-statistic in the two way layout with rank score transformed data. *J. Amer. Statist. Assoc.*, **79**, 668–73.

House, D. E. (1986) A nonparametric version of Williams' test for a randomized block design. *Biometrics*, **42**, 187–90.

Howarth, J. and Curthoys, M. (1987) The political economy of women's higher education in late nineteenth and early twentieth century Britain. *Historical Research*, **60**, 208–31.

Huber, P. J. (1972) Robust statistics; a review. *Ann. Math. Statist.*, **43**, 1041–67.

Huber, P. J. (1977) *Robust Statistical Procedures*, Philadelphia: Society for Industrial and Applied Mathematics.

Hussain, S. S. and Sprent, P. (1983) Non-parametric regression. *J. Roy. Statist. Soc.*, A, **146**, 182–91.

Iman, R. L. and Davenport, J. M. (1980) Approximations of the critical region of the Friedman statistic. *Communications in Statistics*, **A9**, 571–95.

Iman, R. L., Hora, S. C. and Conover, W. J. (1984) Comparison of asymptotically distribution-free procedures for the analysis of complete blocks. *J. Amer. Statist. Soc.*, **79**, 674–85.

Jaeckel, L. A. (1972) Estimating regression coefficients by minimizing the dispersion of residuals. *Ann. Math. Statist.*, **43**, 1449–58.

Jarque, C. M. and Bera, A. K. (1987) A test for normality of observations and regression residuals. *International Statist. Rev.*, **55**, 163–72.

Jarrett, R. G. (1979) A note on the intervals between coal mining disasters. *Biometrika*, **66**, 191–3.

Johnson, R. A., Verrill, S. and Moore, D. H. (1987) Two-sample rank tests for detecting changes that occur in a small proportion of the treated population. *Biometrics*, **43**, 641–55.

Jonckheere, A. R. (1954) A distribution-free *k*-sample test against ordered alternatives. *Biometrika*, **41**, 133–45.

Katti, S. K. (1965) Multivariate covariance analysis. *Biometrics*, **21**, 957–74.

Kendall, M. G. (1938) A new measure of rank correlation. *Biometrika*, **30**, 81–93.

Kendall, M. G. (1962) *Rank Correlation Methods* (3rd edn), London: Griffin.

Kerridge, D. (1975) The interpretation of rank correlations. *Applied Statistics*, **24**, 257.

Kestemont, M-P, (1987) The Kolmogorov distance as comparison measure between parametric and nonparametric Bayesian inference. *The Statistician*, **36**, 259–64.

Kildea, D. G. (1978) Estimators for regression methods – the Brown–Mood approach. Ph.D. Thesis, Latrobe University, Melbourne, Australia.

Kimber, A. C. (1987) When is a χ^2 not a χ^2? *Teaching Statistics*, **9**, 74–7.

Kimura, D. K. and Chikuni, S. (1987) Mixtures of empirical distributions: an iterative application of the age–length key. *Biometrics*, **43**, 23–35.

Klotz, J. (1962) Nonparametric tests for scales. *Ann. Math. Statist.*, **33**, 498–512.

Klotz, J. (1980) A modified Cochran–Friedman test with missing observations and ordered categorical data. *Biometrics*, **36**, 665–70.

Knafl, G., Sacks, J., Spiegelman, C. and Ylvisaker, D. (1984) Nonparametric calibration. *Technometrics*, **26**, 233–41.

Knapp, T. R. (1982) The birthday problem: some empirical data and some approximations. *Teaching Statistics*, **4**, 10–14.

Kolmogorov, A. N. (1933) Sulla determinazione empirica di una legge di distribuzione. *G. Ist. Ital. Attuari*, **4**, 83–91.

Kolmogorov, A. N. (1941) Confidence limits for an unknown distribution function. *Ann. Math. Statist.*, **12**, 461–3.

Kotz, S. and Johnson, N. L. (eds) (1983) *Encyclopedia of Statistical Sciences*, New York: John Wiley & Sons.

Kotze, T. J. v W. and Hawkins, D. M. (1984) The identification of outliers in two way contingency tables using 2×2 subtables. *Applied Statistics*, **33**, 215–23.

Koziol, J. A. (1980) On a Crámer–von Mises-type statistic for testing symmetry. *J. Amer. Statist. Soc.*, **75**, 161–7.

Kruskal, W. H. and Wallis, W. A. (1952) Use of ranks in one-criterion variance analysis. *J. Amer. Statist. Assoc.*, **47**, 583–621.

Lam, F. C. and Longnecker, M. T. (1983) A modified Wilcoxon rank sum test for paired data. *Biometrika*, **70**, 510–13.

Lan, K. K. G. and Wittes, J. T. (1985) Rank tests for survival analysis: a comparison by analogy with games. *Biometrics*, **41**, 1063–9.

Leach, C. (1979) *Introduction to Statistics. A Nonparametric Approach for the Social Sciences*, Chichester: John Wiley & Sons.

Lehmann, E. L. (1975) *Nonparametrics: Statistical Methods Based on Ranks*, San Francisco: Holden Day, Inc.

Lilliefors, H. W. (1967) On the Kolmogorov–Smirnov test for normality with mean and variance unknown. *J. Amer. Statist. Assoc.*, **62**, 399–402.

Lindsey, J. C., Herzberg, A. M. and Watts, D. G. (1987) A method of cluster analysis based on projections and quantile–quantile plots. *Biometrics*, **43**, 327–41.

Lombard, F. (1986) The change point problem for angular data. A nonparametric approach. *Technometrics*, **28**, 391–7.

Lombard, F. (1987) Rank tests for changepoint problems. *Biometrika*, **74**, 615–24.

Lubischew, A. A. (1962) On the use of discriminant functions in taxonomy. *Biometrics*, **18**, 455–77.

McCullagh, P. (1980) Regression models for ordinal data. *J. Roy. Statist. Soc.*, B, **42**, 109–42.

McCullagh, P. and Nelder, J. A. (1983) *Generalized Linear Models*, London: Chapman & Hall.

Mack, G. A. and Wolfe, D. A. (1981) K-sample rank test for umbrella alternatives. *J. Amer. Statist. Assoc.*, **76**, 175–81.

Mann, H. B. and Whitney, D. R. (1947) On a test of whether one of two random variables is stochastically larger than the other. *Ann. Math. Statist.*, **18**, 50–60.

Mantel, N. and Haenszel, W. (1959) Statistical aspects of the analysis of data from retrospective studies of disease. *J. Nat. Cancer Res. Inst.*, **22**, 719–48.

Marascuilo, L. A. and McSweeney, M. (1977) *Nonparametric and Distribution-Free Methods for the Social Sciences*, Monterey: Brooks/Cole Publishing Company.

Marascuilo, L. A. and Serlin, R. C. (1979) Tests and contrasts for comparing change parameters for a multiple sample McNemar data model. *Brit. J. Math. and Statist. Psychol.*, **32**, 105–12.

Maritz, J. S. (1979) On Theil's method in distribution-free regression. *Austral. J. Statist.*, **21**, 30–5.

Maritz, J. S. (1981) *Distribution-Free Statistical Methods*, London: Chapman & Hall.

Markowski, E. P. and Hettmansperger, T. P. (1982) Inferences based on simple rank step scores statistics for the location model. *J. Amer. Statist. Assoc.*, **77**, 901–7.

Mattingley, P. F. (1987) Pattern of horse devolution and tractor diffusion in Illinois, 1920–82. *The Professional Geographer*, **39**, 298–309.

Meddis, R. (1980) Unified analysis of variance by ranks. *Brit. J. Math. and Statist. Psychol.*, **33**, 84–98.

Michalek, J. E. and Mihalko, D. (1984) Linear rank procedures on litter-match models. *Biometrics*, **40**, 487–91.

Mickey, R. M. and Elashoff, R. M. (1985) A generalization of the Mantel–Haenszel estimator of partial association for $2 \times J \times K$ tables. *Biometrics*, **41**, 623–35.

Mood, A. M. (1940) The distribution theory of runs. *Ann. Math. Statist.*, **11**, 367–92.

Mood, A. M. (1954) On the asymptotic efficiency of certain nonparametric two-sample tests. *Ann. Math. Statist.*, **25**, 514–22.

Moses, L. E. (1963) Rank tests for dispersion. *Ann. Math. Statist.*, **34**, 973–83.

Neave, H. R. (1981) *Elementary Statistical Tables*, London: George Allen & Unwin Ltd.

Noether, G. E. (1981) Why Kendall's tau? *Applied Statistics*, **3**, 41–3.

Noether, G. E. (1984) Nonparametrics: the early years – impressions and recollections. *The American Statistician*, **38**, 173–8.

Noether, G. E. (1987a) Sample size determination for some common nonparametric tests. *J. Amer. Statist. Soc.*, **82**, 645–7.

Noether, G. E. (1987b) Mental random numbers: perceived and real randomness. *Teaching Statistics*, **9**, 68–70.

O'Brien, P. C. and Dyck, P. J. (1985) A runs test based on run lengths. *Biometrics*, **41**, 237–44.

Oja, H. (1987) On permutation tests in multiple regression and analysis of covariance problems. *Austral. J. Statist.*, **29**, 91–100.

O'Muircheartaigh, I. G. and Sheil, J. (1983) Fore or five? The indexing of a golf course. *Applied Statistics*, **32**, 287–92.

Page, E. B. (1963) Ordered hypotheses for multiple treatments: a significance test for linear ranks. *J. Amer. Statist. Assoc.*, **58**, 216–30.

Paul, S. R. (1979) Models and estimation procedures for the calibration of examiners. *Brit. J. Math. and Statist. Psychol.*, **32**, 242–51.

Pearson, J. C. G. and Sprent, P. (1968) Trends in hearing loss associated with age or exposure to noise. *Applied Statistics*, **17**, 205–15.

Pearce, S. C. (1965) *Biological Statistics—an Introduction*, New York: McGraw-Hill.

Peto, R. and Peto, J. (1972) Asymptotically efficient rank-invariant test procedures. *J. Roy. Statist. Soc.*, A, **135**, 185–206.

Pettitt, A. N. (1979) A nonparametric approach to the change-point problem. *Applied Statistics*, **28**, 126–35.

Pettitt, A. N. (1981) Posterior probabilities for a change point using ranks. *Biometrika*, **68**, 443–50.

Pettitt, A. N. (1983) Approximate methods using ranks for regression with censored data. *Biometrika*, **70**, 121–32.

Pettitt, A. N. (1985) Re-weighted least squares estimation with censored and grouped data: an application of the EM algorithm. *J. Roy. Statist. Soc.*, B, **47**, 253–60.

Pitman, E. J. G. (1937) Significance tests that may be applied to samples from any population. *J. Roy. Statist. Soc., Suppt.*, **4**, 119–130.

Pitman, E. J. G. (1938) Significance tests that may be applied to samples from any population. III. The analysis of variance test. *Biometrika*, **29**, 322–35.

Pitman, E. J. G. (1948) Mimeographed lecture notes on nonparametric statistics. Columbia University.

Plackett, R. L. (1981) *The Analysis of Categorical Data* (2nd edn), London: Charles Griffin & Co.

Prentice, R. L. (1978) Linear rank tests with right censored data. *Biometrika*, **65**, 167–79.

Quade, D. (1979) Using weighted rankings in the analysis of complete blocks with additive block effects. *J. Amer. Statist. Assoc.*, **74**, 680–3.

Quade, D. (1982) Nonparametric analysis of covariance by matching. *Biometrics*, **38**, 597–611.

Quenouille (1949) Approximate tests of correlation in time series. *J. Roy. Statist. Soc.*, B, **11**, 18–84.

Ramsay, F. L. (1972) A Bayesian approach to bioassay. *Biometrics*, **28**, 841–58.

Randles, R. H. and Wolfe, D. A. (1979) *Introduction to the Theory of Nonparametric Statistics*, New York: John Wiley & Sons.

Robinson, J. A. (1983) Bootstrap confidence intervals in location-scale models with progressive censoring. *Technometrics*, **25**, 179–87.

Rogerson, P. A. (1987) Changes in US national mobility levels. *The Professional Geographer*, **39**, 344–51.

Rosenthal, I. and Ferguson, T. S. (1965) An asymptotic distribution-free multiple comparison method with application to the problem of n rankings of m objects. *Brit. J. Math. and Statist. Psychol.*, **18**, 243–54.

Rowntree, D. (1981) *Statistics Without Tears*, London: Penguin Books.

Sandford, M. D. (1985) Nonparametric one-sided confidence intervals for an unknown distribution function using censored data. *Technometrics*, **27**, 41–8.

Saunders, R. and Laud, P. (1980) The multidimensional Kolmogorov goodness of fit test. *Biometrika*, **67**, 237.

Scholz, F. W. (1978) Weighted median regression estimates. *Annals of Statistics*, **6**, 603–9.

Scott, A. J., Smith, T. M. F. and Jones, R. G. (1977) The application of time series methods to the analysis of repeated surveys. *Internat. Statist. Rev.*, **45**, 13–28.

Sen, P. K. (1968) Estimates of the regression coefficient based on Kendall's tau. *J. Amer. Statist. Assoc.*, **63**, 1379–89.

Shirley, E. (1977) A nonparametric equivalent of Williams' test for contrasting existing dose levels of a treatment. *Biometrics*, **33**, 386–9.

Shirley, E. A. C. (1987) Applications of ranking methods to multiple comparison procedures and factorial experiments. *Applied Statistics*, **36**, 205–13.

Siegel, S. (1956) *Nonparametric Statistics for the Behavioral Sciences*, New York: McGraw-Hill.

Sievers, G. L. (1978) Weighted rank statistics for simple linear regression. *J. Amer. Statist. Assoc.*, **73**, 628–31.

Silverman, B. W. (1984) A fast and efficient cross-validation method for smoothing parameter choice in spline regression. *J. Amer. Statist. Assoc.*, **79**, 584–9.

Silverman, B. W. (1985) Some aspects of the spline smoothing approach to nonparametric regression curve fitting. *J. Roy. Statist. Soc.*, B, **47**, 1–52.

Silverman, B. W. and Wood, J. T. (1987) The nonparametric estimation of branching curves. *J. Amer. Statist. Assoc.*, **82**, 551–8.

Silverman, B. W. and Young, G. A. (1987) The Bootstrap: to smooth or not to smooth? *Biometrika*, **74**, 469–79.

Simonoff, J. S., Hochberg, Y. and Reiser, B. (1986) Alternative estimation procedures for $\Pr(X < Y)$ in categorized data. *Biometrics*, **42**, 897–907.

Singer, B. (1979) Distribution-free methods for non-parametric problems. A classified and selected bibliography. *Brit. J. Math. and Statist. Psychol.*, **32**, 1–60.

Slud, E. V., Byar, D. P. and Green, S. B. (1984) A comparison of reflected versus test-based confidence intervals for the median survival time based on censored data. *Biometrics*, **40**, 587–600.

Smirnov, N. V. (1939) On the estimation of discrepancy between empirical curves of distribution for two independent samples (in Russian). *Bulletin Moscow University*, **2**, 3–16.

Smirnov, N. V. (1948) Tables for estimating the goodness of fit of empirical distributions. *Ann. Math. Statist.*, **19**, 279–81.

Smith, E. P. and van Belle, G. (1984) Nonparametric estimation of species richness. *Biometrics*, **40**, 119–29.

Smith, R. L. and Naylor, J. C. (1987) A comparison of likelihood and Bayesian estimators of the three-parameter Weibull distribution. *Applied Statistics*, **36**, 358–69.

Snee, R. D. (1985) Graphical display of results of three treatment randomized block experiments. *Applied Statistics*, **34**, 71–7.

Spearman, C. (1904) The proof and measurement of association between two things. *Amer. J. Psychol.*, **15**, 72–101.

Sprent, P. (1977) *Statistics in Action*, London: Penguin Books.

Sprent, P. (1981) *Quick Statistics*, London: Penguin Books.

Spurrier, J. D. and Hewett, J. E. (1980) Two-stage test of independence using Kendall's statistic. *Biometrics*, **36**, 517–22.

Sweeting, T. J. (1982) A Bayesian analysis of some pharmacological data using a random coefficient regression model. *Applied Statistics*, **31**, 205–13.

Theil, H. (1950) A rank invariant method of linear and polynomial regression analysis, I, II, III. *Proc. Kon. Nederl. Akad. Wetensch., A*, **53**, 386–92, 521–5, 1397–412.

Titterington, D. M. (1983) Minimum distance non-parametric estimation of mixture proportions. *J. Roy. Statist. Soc.*, B, **45**, 37–46.

Tukey, T. W. (1958) Bias and confidence is not quite large samples. Abstract. *Ann. Math. Statist.*, **29**, 614.

Turnbull, B. W. and Mitchell, T. J. (1984) Nonparametric estimation of the distribution of time to onset for specific diseases in survival/sacrifice experiments. *Biometrics*, **40**, 41–50.

Wahrendorf, J., Becher, H. and Brown, C. C. (1987) Bootstrap comparisons of non-nested generalized linear models: applications in survival analysis and epidemiology. *Applied Statistics*, **36**, 72–81.

Wald, A. and Wolfowitz, J. (1940) On a test whether two samples are from the same population. *Ann. Math. Statist.*, **11**, 147–62.

Wegman, E. J. and Wright, I. W. (1983) Splines in statistics. *J. Amer. Statist. Assoc.*, **78**, 351–65.

Wei, L. J. (1982) Asymptotically distribution-free simultaneous confidence regions for treatment differences in a randomized complete block design. *J. Roy. Statist. Soc.*, B, **44**, 201–8.

Wilcoxon, F. (1945) Individual comparisons by ranking methods. *Biometrics*, **1**, 80–3.

Willemain, T. R. (1980) Estimating the population median by nomination sampling. *J. Amer. Statist. Assoc.*, **75**, 908–11.

Williams, D. A. (1986) A note on Shirley's nonparametric test for comparing several dose levels with a zero-dose control. *Biometrics*, **42**, 183–6.

Woolson, R. F. (1981) Rank tests and a one-sample log rank test for comparing observed survival data to a standard population. *Biometrics*, **37**, 687–96.

Woolson, R. F. and Lachenbruch, P. A. (1980) Rank tests for censored matched pairs. *Biometrika*, **67**, 597–606.

Woolson, R. F. and Lachenbruch, P. A. (1981) Rank tests for censored randomized block designs. *Biometrika*, **68**, 427–35.

Woolson, R. F. and Lachenbruch, P. A. (1983) Rank analysis of covariance with right-censored data. *Biometrics*, **39**, 727–33.

Yates, F. (1984) Tests of significance for 2 × 2 contingency tables. *J. Roy. Statist. Soc.*, A, **147**, 426–63.

Solutions to odd-numbered exercises

Most exercises in this book are open-ended in that they require, as do most practical problems in statistics, not simply numerical outcomes (a statistic, significance level, confidence interval, etc.) but an interpretation of such numerical values in the context of the real-world problem giving rise to the original data. The brief summaries below are not solutions in this sense but rather a guide to indicate appropriate calculations. In many cases there are alternative tests or estimation procedures that could well have been used; these will often, but not invariably, lead to similar conclusions. Bearing such points in mind it is hoped this section will be an aid to the reader having difficulties with any of the odd-numbered exercises.

Chapter 1

1.1. Reject $H_0: \theta = 460$ outside confidence intervals based on sign or t-tests.

1.5. 4/165.

Chapter 2

2.1. Sign test appropriate; reject $H_0: \theta = 422$ at 5% level in one- but not in two-tail test (although actual level only 5.8%). Inflation, increasing traffic density may justify one-tail test. 95% confidence interval (422, 1770).

2.3. 315.75.

2.5. Do not reject $H_0: \theta = 5$ at 5% level with sign test; reject with Wilcoxon using large-sample approximation allowing for ties. 95% confidence interval assuming symmetry (2.5, 4.5); without (2, 6). Data suggest symmetry assumption reasonable.

2.7. Sign test rejects H_0 at 5% level.

2.9. True median 300.5. In long run true median not expected to be in 95% confidence interval for about 5% of samples.

2.11. Do not reject $H_0: p = \frac{1}{2}$ for sample of 18. For larger sample reject H_0 at 0.1% level; power increases with sample size.

2.13. Suggestion of asymmetry (6 at 77); 95% confidence limit based on large-sample sign test (69, 73); claim of 'almost certainty' rather too sweeping.

2.15. Sign test clearly rejects H_0. (Wilcoxon probably justified, but unnecessary for question asked here.)

2.17. Cox–Stuart test just fails to detect trend in two-tail test (one-tail not justified for question as formulated). Note 'actual' level is closest actual level to nominal 5%.

Chapter 3

3.1. Lilliefors statistic 0.253; significant at 5% and almost at 1% level.

3.3. $z = -2.21$. Reject H_0 at 5% level. What score did you assign to zero difference?

3.5. Negative sum of 10 can be achieved in three ways only. Reject in two-tail test at 5% level (actual level 6/126).

3.7. Data may come from given distribution. Kolmogorov statistic 0.268; not significant.

3.9. Both tests significant at 5% and almost at 1% level.

Chapter 4

4.1. One-tail sign test does not reject; Wilcoxon rejects at 5% level.

4.3. Reject H_0: no difference, at 5% level.

4.5. Do not reject H_0: no influence, with sign test. Some doubt about symmetry, but Wilcoxon one-tail significant at 5% level; one-tail test justified if we assume paranormal influence, if any, will be positive.

4.7. No.

4.9. No.

4.11. Neither sign nor Wilcoxon test reject hypothesis. Two-tail test appropriate because although poor weather will almost certainly increase median scores it may do so by more or less than 3, which is only commentator's (perhaps expert) guess.

4.13. $z = 2.01$; reject hypothesis of no difference at 5% level.

Chapter 5

5.3. $U = 13$, exceeds critical value of 12 needed in two-tail test.

5.5. $U = 275$ for men, $U = 116$ for women; $z = -2.22$. Significant at 5% level. Men show greater anxiety.

5.7. Smirnov statistic 0.417; not significant.

5.9. (i) $U = 35$, not significant; (ii) squared ranks or parametric F-test indicate variance difference. Did you try Siegel–Tukey? (iii) Smirnov test not significant, but Cramer–von Mises indicates significance at 5% level; (iv) Lilliefors test does not reject normality for either set.

5.11. Yes, $U = 12$.

5.13. $U = 17$, reject H_0 at 0.1% level using normal approximation. For confidence limits reject 62 largest/smallest differences (normal approximation based on (5.3) with $z = -1.96$, $m = 9$, $n = 25$ gives 61). 95% confidence interval is (5, 13). Hodges–Lehmann estimator is 9. The 95% interval based on t-distribution is (5.07, 12.95).

Chapter 6

6.1. $T = 10.76$, 2 d.f. Highly significant. Least significance difference test implies recall rates significantly lower for second and third groups than for first.

6.3. Not significant using ranks or scores. $T = 3.37$ for Kruskal–Wallis.

6.5. $T = 9.06$ for Kruskal–Wallis. Significant difference; Vulliamy uses longest sentences (fewest per page).

6.9. (i) Not significant, $T = 1.42$. (ii) Significant at 1% level, $T = 5.28$ using (6.6).

6.11. Not significant, $T = 4.70$ using (6.1) to allow for ties.

Chapter 7

7.1. Do not reject H_0: no trend.

7.3. $\rho = 0$, $\tau = 1/21$. Strong evidence against association.

7.5. $\rho = 0.83$, $\tau = 0.667$. ρ significant at 5% level but τ just short of value required for significance (would have attained significance if tie had been a concordance).

7.7. $y = -13.6 + 2.42x$. Graph shows nonlinearity. A good idea to plot scatter diagram first to save trouble of fitting inappropriate line. Somewhat caddish to word question the way I did, but people all too often go through formal analyses without thinking whether they are justified.

7.9. For equally spaced points all denominators in the abbreviated Theil method are equal; estimators are effectively Cox–Stuart differences each divided by same constant, whence claim follows.

7.11. (i) $T = 2.747$ (equation (6.6)); significant at 5% level. Clearly examiners show some agreement but this is not consistent across all candidates; (ii) $T = 4.416$; significant at 1% level; in particular, examiner 5 consistently awards high marks.

7.13. $y = 5.74 + 0.0064x$. 95% confidence interval $(0.0032, 0.0095)$

Chapter 8

8.1. No, $T = 2.37$.

8.3. $T = 5.31$, 2 d.f. Not quite significant at 5% level. A larger sample might clarify.

8.5. $T = 60.22$. Very highly significant; low incomes appear to reduce visits.

8.7. Highly significant indication that manufacturer's claim unjustified. *Hint*: Under H_0 the probabilities that 0, 1, 2, 3 or 4 parts survive has binomial distribution with $p = 0.95$.

8.9. Yes. We would expect 36.8% positive responses (estimated from data) under H_0.

8.11. Overwhelming evidence death times not spread uniformly.

8.13. $T = 17.38$; highly significant. Strong evidence that temperatures affect output.

8.15. $T = 23$ (McNemar). Very highly significant indication of changes in attitude.

8.17. $T < 1$. First-order interaction model adequate, $\hat{e}_{111} = 8.46$.

8.19. Overwhelming evidence of non-randomness; a tendency for too few pairs 11, 22, 33 reflecting instinct to avoid an obvious pattern. Note that expected numbers in each cell are 50; they should not be based on marginal totals; test is a goodness of fit test, not one of category independence.

Chapter 9

9.3. mean 0, unbiased; variance 12.5, bias in sample estimate is -2.5.

9.7. 0.0018.

Index